higher-level thinking Questions
Biology

questions by
Angela Manzi
Michael Michels

created and designed by
Miguel Kagan

layout by
Miles Richey

illustrated by
Erin Kant and
Celso Rodriguez

Kagan

Kagan

© 2005 by **Kagan Publishing**

This book is published and distributed by **Kagan Publishing**. All rights are reserved by **Kagan Publishing**. No part of this publication may be reproduced or transmitted in any form by any means, electronic or mechanical, including photocopy, recording, or any information storage and retrieval system, without prior written permission from **Kagan Publishing**. The blackline masters included in this book may be duplicated only by classroom teachers who purchase the book, and only for use in their own classrooms. To obtain additional copies of this book, other **Kagan Publishing** publications, or information regarding **Kagan Publishing** professional development, contact **Kagan Publishing.**

Kagan Publishing

981 Calle Amanecer

San Clemente, CA 92673

(949) 545-6300

1 (800) 933-2667

www.KaganOnline.com

ISBN: 978-1-879097-85-8

Table of Contents

Introduction 3

1. Animals 27
2. Biochemistry 35
3. Bioenergetics 43
4. Biotechnology and Ethics 51
5. Body Systems 59
6. Cells.................... 67
7. Classification........ 75
8. Ecology 83
9. Evolution.............. 91
10. Fungi 99
11. Genetics............. 107
12. Methods and Tools 115
13. Monera.............. 123
14. Plants 131
15. Protista 139
16. Viruses and Diseases...... 147

> **I had six honest serving men They taught me all I knew: Their names were Where and What and When and Why and How and Who.**
>
> — Rudyard Kipling

Introduction

In your hands you hold a powerful book. It is a member of a series of transformative blackline activity books. Between the covers, you will find questions, questions, and more questions! But these are no ordinary questions. These are the important kind—higher-level thinking questions—the kind that stretch your students' minds; the kind that release your students' natural curiosity about the world; the kind that rack your students' brains; the kind that instill in your students a sense of wonderment about your curriculum.

But we are getting a bit ahead of ourselves. Let's start from the beginning. Since this is a book of questions, it seems only appropriate for this introduction to pose a few questions—about the book and its underlying educational philosophy. So Mr. Kipling's Six Honest Serving Men, if you will, please lead the way:

What? What are higher-level thinking questions?

This is a loaded question (as should be all good questions). Using our analytic thinking skills, let's break this question down into two smaller questions: 1) What is higher-level thinking? and 2) What are questions? When we understand the types of thinking skills and the types of questions, we can combine the best of both worlds, crafting beautiful questions to generate the range of higher-level thinking in our students!

Types of Thinking

There are many different types of thinking. Some types of thinking include:

- **applying**
- **associating**
- **comparing**
- **contrasting**
- **defining**
- **elaborating**
- **empathizing**
- **experimenting**
- **generalizing**
- **investigating**
- **making analogies**
- **planning**
- **prioritizing**
- **recalling**
- **reflecting**
- **reversing**
- **sequencing**
- **summarizing**
- **synthesizing**
- **assessing**
- **augmenting**
- **connecting**
- **decision-making**
- **drawing conclusions**
- **eliminating**
- **evaluating**
- **explaining**
- **inferring consequences**
- **inventing**
- **memorizing**
- **predicting**
- **problem-solving**
- **reducing**
- **relating**
- **role-taking**
- **substituting**
- **symbolizing**
- **understanding**
- **thinking about thinking (metacognition)**

This is quite a formidable list. It's nowhere near complete. Thinking is a big, multifaceted phenomenon. Perhaps the most widely recognized system for classifying thinking and classroom questions is Benjamin Bloom's Taxonomy of Thinking Skills. Bloom's Taxonomy classifies thinking skills into six hierarchical levels. It begins with the lower levels of thinking skills and moves up to higher-level thinking skills: 1) Knowledge, 2) Comprehension, 3) Application, 4) Analysis, 5) Synthesis, 6) Evaluation. See Bloom's Taxonomy on the following page.

Bloom's Taxonomy

Higher-Level Thinking
- Evaluation
- Synthesis
- Analysis
- Application
- Comprehension
- Knowledge

Lower-Level Thinking

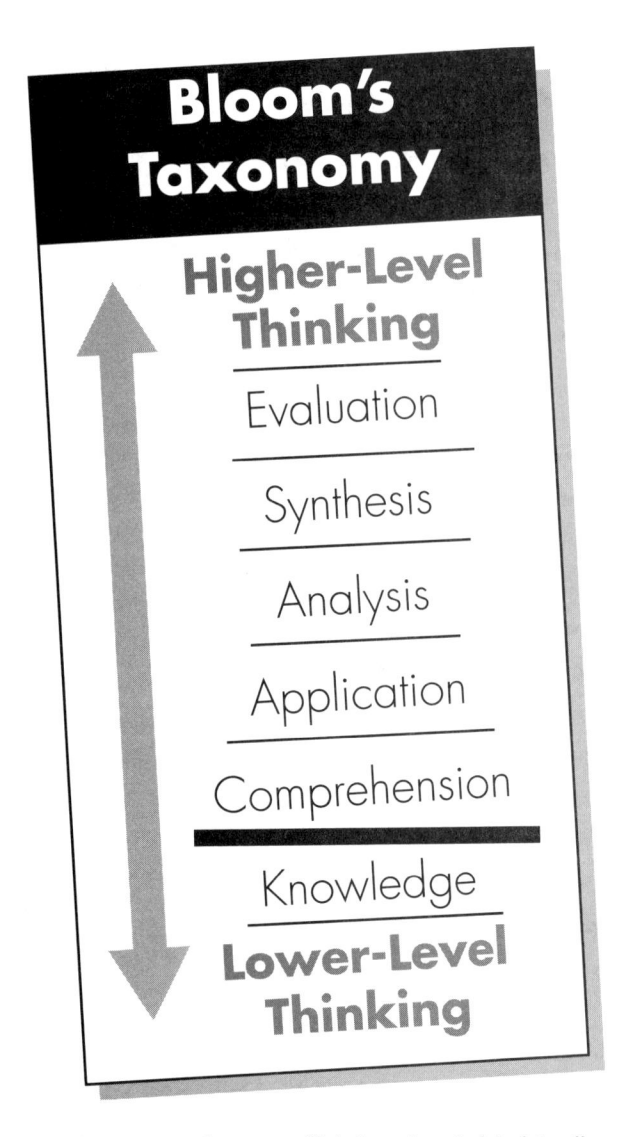

In education, the term "higher-level thinking" often refers to the higher levels of Mr. Bloom's taxonomy. But Bloom's Taxonomy is but one way of organizing and conceptualizing the various types of thinking skills.

There are many ways we can cut the thinking skills pie. We can alternatively view the many different types of thinking skills as, well…many different skills. Some thinking skills may be hierarchical. Some may be interrelated. And some may be relatively independent.

In this book, we take a pragmatic, functional approach. Each type of thinking skill serves a different function. So called "lower-level" thinking skills are very useful for certain purposes. Memorizing and understanding information are invaluable skills that our students will use throughout their lives. But so too are many of the "higher-level" thinking skills on our list. The more facets of students' thinking skills we develop, the better we prepare them for lifelong success.

Because so much classroom learning heretofore has focused on the "lower rungs" of the thinking skills ladder—knowledge and comprehension, or memorization and understanding—in this series of books we have chosen to focus on questions to generate "higher-level" thinking. This book is an attempt to correct the imbalance in the types of thinking skills developed by classroom questions.

Types of Questions

As we ask questions of our students, we further promote cognitive development when we use Fat questions, Low-Consensus questions, and True questions.

Fat Questions vs. Skinny Questions

Skinny questions are questions that require a skinny answer. For example, after reading a poem, we can ask: "Did you like the poem?" Even though this question could be categorized as an Evaluation question—Bloom's highest level of thinking— it can be answered with one monosyllabic word: "Yes" or "No." How much thinking are we actually generating in our students?

We can reframe this question to make it a fat question: "What things did you like about the poem? What things did you dislike?" Notice no short answer will do. Answering this fattened-up question requires more elaboration. These fat questions presuppose not that there is only one thing but things plural that the student liked and things that she did not like. Making things plural is one way to make skinny questions fat. Students stretch their minds to come up with multiple ideas or solutions. Other easy ways to make questions fat is to add "Why or why not?" or "Explain" or "Describe" or "Defend your position" to the end of a question. These additions promote elaboration beyond a skinny answer. Because language and thought are intimately intertwined, questions that require elaborate responses stretch students' thinking: They grapple to articulate their thoughts.

The type of questions we ask impact not just the type of thinking we develop in our students, but also the depth of thought. Fat questions elicit fat responses. Fat responses develop both depth of thinking and range of thinking skills. The questions in this book are designed to elicit fat responses—deep and varied thinking.

High-Consensus Questions vs. Low-Consensus Questions

A high-consensus question is one to which most people would give the same response, usually a right or wrong answer. After learning about sound, we can ask our students: "What is the name of a room specially designed to improve acoustics for the audience?" This is a high-consensus question. The answer (auditorium) is either correct or incorrect.

Compare the previous question with a low-consensus question: "If you were going to build an auditorium, what special design features would you take into consideration?" Notice, to the low-consensus question there is no right or wrong answer. Each person formulates his or her unique response. To answer, students must apply what they learned, use their ingenuity and creativity.

High-consensus questions promote convergent thinking. With high-consensus questions we strive to direct students *what to think*. Low-consensus questions promote divergent thinking, both critical and creative. With low-consensus questions we strive to develop students' *ability to think*. The questions in this book are low-consensus questions designed to promote independent, critical and creative thought.

True Questions vs. Review Questions

We all know what review questions are. They're the ones in the back of every chapter and unit. Review questions ask students to regurgitate previously stated or learned information. For example, after learning about the rain forest we may ask: "What percent of the world's oxygen does the rain forest produce?" Students can go back a few pages in their books or into their memory banks and pull out the answer. This is great if we are working on memorization skills, but does little to develop "higher-order" thinking skills.

True questions, on the other hand, are meaningful questions—questions to which we do not know the answer. For example: "What might happen if all the world's rain forests were cut down?" This is a hypothetical; we don't know the answer but considering the question forces us to think. We infer some logical consequences based on what we know. The goal of true questions is not a correct answer, but the thinking journey students take to create a meaningful response. True questions are more representative of real life. Seldom is there a black and white answer. In life, we struggle with ambiguity, confounding variables, and uncertain outcomes. There are millions of shades of gray. True questions prepare students to deal with life's uncertainties.

When we ask a review question, we know the answer and are checking to see if the student does also. When we ask a true question, it is truly a question. We don't necessarily know the answer and neither does the student. True questions are

> **Education is not the filling of a pail, but the lighting of a fire.**
> — William Butler Yeats

Types of Questions

Skinny → **Fat**
- Short Answer
- Shallow Thinking

- Elaborated Answer
- Deep Thinking

High-Consensus → **Low-Consensus**
- Right or Wrong Answer
- Develops Convergent Thinking
- "What" to Think

- No Single Correct Answer
- Develops Divergent Thinking
- "How" to Think

Review → **True**
- Asker Knows Answer
- Checking for Correctness

- Asker Doesn't Know Answer
- Invitation to Think

---

often an invitation to think, ponder, speculate, and engage in a questioning process.

We can use true questions in the classroom to make our curriculum more personally meaningful, to promote investigation, and awaken students' sense of awe and wonderment in what we teach. Many questions you will find in this book are true questions designed to make the content provocative, intriguing, and personally relevant.

The box above summarizes the different types of questions. The questions you will find in this book are a move away from skinny, high-consensus, review questions toward fat, low-consensus true questions. As we ask these types of questions in our class, we transform even mundane content into a springboard for higher-level thinking. As we integrate these question gems into our daily lessons, we create powerful learning experiences. ***We do not fill our students' pails with knowledge; we kindle their fires to become lifetime thinkers.***

Why? Why should I use higher-level thinking questions in my classroom?

As we enter the new millennium, major shifts in our economic structure are changing the ways we work and live. The direction is increasingly toward an information-based, high-tech economy. The sum of our technological information is exploding. We could give you a figure how rapidly information is doubling, but by the time you read this, the number would be outdated! No kidding.

But this is no surprise. This is our daily reality. We see it around us everyday and on the news: cloning, gene manipulation, e-mail, the Internet, Mars rovers, electric cars, hybrids, laser surgery, CD-ROMs, DVDs. All around us we see the wheels of progress turning: New discoveries, new technologies, a new societal knowledge and information base. New jobs are being created today in fields that simply didn't exist yesterday.

How do we best prepare our students for this uncertain future—a future in which the only constant will be change? As we are propelled into a world of ever-increasing change, what is the relative value of teaching students facts versus thinking skills? This point becomes even more salient when we realize that students cannot master everything, and many facts will soon become obsolete. Facts become outdated or irrelevant. Thinking skills are for a lifetime. Increasingly, how we define educational success will be away from the quantity of information mastered. Instead, we will define success as our students' ability to generate questions, apply, synthesize, predict, evaluate, compare, categorize.

If we as a professionals are to proactively respond to these societal shifts, thinking skills will become central to our curriculum. Whether we teach thinking skills directly, or we integrate them into our curriculum, the power to think is the greatest gift we can give our students!

We believe the questions you will find in this book are a step in the direction of preparing students for lifelong success. The goal is to develop independent thinkers who are critical and creative, regardless of the content. We hope the books in this series are more than sets of questions. We provide them as a model approach to questioning in the classroom.

On pages 8 and 9, you will find Questions to Engage Students' Thinking Skills. These pages contain numerous types of thinking and questions designed to engage each thinking skill. As you make your own questions for your students with your own content, use these question starters to help you frame your questions to stimulate various facets of your students' thinking skills. Also let your students use these question starters to generate their own higher-level thinking questions about the curriculum.

> **Virtually the only predictable trend is continuing change.**
> — Dr. Linda Tsantis, Creating the Future

Who?
Who is this book for?

This book is for you and your students, but mostly for your students. It is designed to help make your job easier. Inside you will find hundreds of ready-to-use reproducible questions. Sometimes in the press for time we opt for what is easy over what is best. These books attempt to make easy what is best. In this treasure chest, you will find hours and hours of timesaving ready-made questions and activities.

Place Higher-Level Thinking In Your Students' Hands

As previously mentioned, this book is even more for your students than for you. As teachers, we ask a tremendous number of questions. Primary teachers ask 3.5 to 6.5 questions per minute! Elementary teachers average 348 questions a day. How many questions would you predict our students ask? Researchers asked this question. What they found was shocking: Typical students ask approximately one question per month.* One question per month!

Although this study may not be representative of your classroom, it does suggest that in general, as teachers we are missing out on a very powerful force—student-generated questions. The capacity to answer higher-level thinking questions is

---
* Myra & David Sadker, "Questioning Skills" in *Classroom Teaching Skills*, 2nd ed. Lexington, MA: D.C. Heath & Co., 1982.

Questions to Engage Students' Thinking Skills

Analyzing
- How could you break down…?
- What components…?
- What qualities/characteristics…?

Applying
- How is _____ an example of…?
- What practical applications…?
- What examples…?
- How could you use…?
- How does this apply to…?
- In your life, how would you apply…?

Assessing
- By what criteria would you assess…?
- What grade would you give…?
- How could you improve…?

Augmenting/Elaborating
- What ideas might you add to…?
- What more can you say about…?

Categorizing/Classifying/Organizing
- How might you classify…?
- If you were going to categorize…?

Comparing/Contrasting
- How would you compare…?
- What similarities…?
- What are the differences between…?
- How is _____ different…?

Connecting/Associating
- What do you already know about…?
- What connections can you make between…?
- What things do you think of when you think of…?

Decision-Making
- How would you decide…?
- If you had to choose between…?

Defining
- How would you define…?
- In your own words, what is…?

Describing/Summarizing
- How could you describe/summarize…?
- If you were a reporter, how would you describe…?

Determining Cause/Effect
- What is the cause of…?
- How does _____ effect _____?
- What impact might…?

Drawing Conclusions/Inferring Consequences
- What conclusions can you draw from…?
- What would happen if…?
- What would have happened if…?
- If you changed _____, what might happen?

Eliminating
- What part of _____ might you eliminate?
- How could you get rid of…?

Evaluating
- What is your opinion about…?
- Do you prefer…?
- Would you rather…?
- What is your favorite…?
- Do you agree or disagree…?
- What are the positive and negative aspects of…?
- What are the advantages and disadvantages…?
- If you were a judge…?
- On a scale of 1 to 10, how would you rate…?
- What is the most important…?
- Is it better or worse…?

Explaining
- How can you explain…?
- What factors might explain…?

Experimenting
- How could you test…?
- What experiment could you do to…?

Generalizing
- What general rule can…?
- What principle could you apply…?
- What can you say about all…?

Interpreting
- Why is ____ important?
- What is the significance of…?
- What role…?
- What is the moral of…?

Inventing
- What could you invent to…?
- What machine could…?

Investigating
- How could you find out more about…?
- If you wanted to know about…?

Making Analogies
- How is ____ like ____?
- What analogy can you invent for…?

Observing
- What observations did you make about…?
- What changes…?

Patterning
- What patterns can you find…?
- How would you describe the organization of…?

Planning
- What preparations would you…?

Predicting/Hypothesizing
- What would you predict…?
- What is your theory about…?
- If you were going to guess…?

Prioritizing
- What is more important…?
- How might you prioritize…?

Problem-Solving
- How would you approach the problem?
- What are some possible solutions to…?

Reducing/Simplifying
- In a word, how would you describe…?
- How can you simplify…?

Reflecting/Metacognition
- What would you think if…?
- How can you describe what you were thinking when…?

Relating
- How is ____ related to ____?
- What is the relationship between…?
- How does ____ depend on ____?

Reversing/Inversing
- What is the opposite of…?

Role-Taking/Empathizing
- If you were (someone/something else)…?
- How would you feel if…?

Sequencing
- How could you sequence…?
- What steps are involved in…?

Substituting
- What could have been used instead of…?
- What else could you use for…?
- What might you substitute for…?
- What is another way…?

Symbolizing
- How could you draw…?
- What symbol best represents…?

Synthesizing
- How could you combine…?
- What could you put together…?

a wonderful skill we can give our students, as is the skill to solve problems. Arguably more important skills are the ability to find problems to solve and formulate questions to answer. If we look at the great thinkers of the world—the Einsteins, the Edisons, the Freuds—their thinking is marked by a yearning to solve tremendous questions and problems. It is this questioning process that distinguishes those who illuminate and create our world from those who merely accept it.

Make Learning an Interactive Process

Higher-level thinking is not just something that occurs between students' ears! Students benefit from an interactive process. This basic premise underlies the majority of activities you will find in this book.

As students discuss questions and listen to others, they are confronted with differing perspectives and are pushed to articulate their own thinking well beyond the level they could attain on their own. Students too have an enormous capacity to mediate each other's learning. When we heterogeneously group students to work together, we create an environment to move students through their zone of proximal development. We also provide opportunities for tutoring and leadership. Verbal interaction with peers in cooperative groups adds a dimension to questions not available with whole-class questions and answers.

> **Asking a good question requires students to think harder than giving a good answer.**
> — Robert Fisher, Teaching Children to Learn

Reflect on this analogy: If we wanted to teach our students to catch and throw, we could bring in one tennis ball and take turns throwing it to each student and having them throw it back to us. Alternatively, we could bring in twenty balls and have our students form small groups and have them toss the ball back and forth to each other. Picture the two classrooms: One with twenty balls being caught at any one moment, and the other with just one. In which class would students better and more quickly learn to catch and throw?

The same is true with thinking skills. When we make our students more active participants in the learning process, they are given dramatically more opportunities to produce their own thought and to strengthen their own thinking skills. Would you rather have one question being asked and answered at any one moment in your class, or twenty? Small groups mean more questioning and more thinking. Instead of rarely answering a teacher question or rarely generating their own question, asking and answering questions becomes a regular part of your students' day. It is through cooperative interaction that we truly turn our classroom into a higher-level think tank. The associated personal and social benefits are invaluable.

When?
When do I use higher-level thinking questions?

Do I use these questions at the beginning of the lesson, during the lesson, or after? The answer, of course, is all of the above.

Use these questions or your own thinking questions at the beginning of the lesson to provide a motivational set for the lesson. Pique students' interest about the content with some provocative questions: "What would happen if we didn't have gravity?" "Why did Pilgrims get along with some Native Americans, but not others?" "What do you think this book will be about?" Make the content personally relevant by bringing in students' own knowledge, experiences, and feelings about the content: "What do you know about spiders?" "What things do you like about mystery stories?" "How would you feel if explorers invaded your land and killed your family?" "What do you wonder about electricity?"

Use the higher-level thinking questions throughout your lessons. Use the many questions and activities in this book not as a replacement of your curriculum, but as an additional avenue to explore the content and stretch students' thinking skills.

Use the questions after your lesson. Use the higher-level thinking questions, a journal writing activity, or the question starters as an extension activity to your lesson or unit.

Or just use the questions as stand-alone sponge activities for students or teams who have finished their work and need a challenging project to work on.

It doesn't matter when you use them, just use them frequently. As questioning becomes a habitual part of the classroom day, students' fear of asking silly questions is diminished. As the ancient Chinese proverb states, "Those who ask a silly question may seem a fool for five minutes, but those who do not ask remain a fool for life."

> **The important thing is to never stop questioning.**
> — Albert Einstein

As teachers, we should make a conscious effort to ensure that a portion of the many questions we ask on a daily basis are those that move our students beyond rote memorization. When we integrate higher-level thinking questions into our daily lessons, we transform our role from transmitters of knowledge to engineers of learning.

Where?
Where should I keep this book?

Keep it close by. Inside there are 16 sets of questions. Pull it out any time you teach these topics or need a quick, easy, fun activity or journal writing topic.

How?
How do I get the most out of this book?

In this book you will find 16 topics arranged alphabetically. For each topic there are reproducible pages for: 1) 16 Question Cards, 2) a Journal Writing activity page, 3) and a Question Starters activity page.

1. Question Cards

The Question Cards are truly the heart of this book. There are numerous ways the Question Cards can be used. After the other activity pages are introduced, you will find a description of a variety of engaging formats to use the Question Cards.

Specific and General Questions

Some of the questions provided in this book series are content-specific and others are content-free. For example, the literature questions in the Literature books are content-specific. Questions for the Great Kapok Tree deal specifically with that literature selection. Some language arts questions in the Language Arts book, on the other hand, are content-free. They are general questions that can be used over and over again with new content. For example, the Book Review questions can be used after reading any book. The Story Structure questions can be used after reading any story. You can tell by glancing at the title of the set and some of the questions whether the set is content-specific or content-free.

A Little Disclaimer

Not all of the "questions" on the Question Cards are actually questions. Some instruct students to do something. For example, "Compare and contrast…" We can also use these directives to develop the various facets of students' thinking skills.

The Power of Think Time

As you and your students use these questions, don't forget about the power of Think Time! There are two different think times. The first is the time between the question and the response. The second is the time between the response and feedback on the response. Think time has been shown to greatly enhance the quality of student thinking. If students are not pausing for either think time, or doing it too briefly, emphasize its importance. Five little seconds of silent think time after the question and five more seconds before feedback are proven, powerful ways to promote higher-level thinking in your class.

Use Your Question Cards for Years

For attractive Question Cards that will last for years, photocopy them on color card-stock paper and laminate them. To save time, have the Materials Monitor from each team pick up one card set, a pair of scissors for the team, and an envelope or rubber band. Each team cuts out their own set of Question Cards. When they are done with the activity, students can place the Question Cards in the envelope and write the name of the set on the envelope or wrap the cards with a rubber band for storage.

2. Journal Question

The Journal Writing page contains one of the 16 questions as a journal writing prompt. You can substitute any question, or use one of your own. The power of journal writing cannot be overstated. The act of writing takes longer than speaking and thinking. It allows the brain time to make deep connections to the content. Writing requires the writer to present his or her response in a clear, concise language. Writing develops both strong thinking and communication skills.

A helpful activity before journal writing is to have students discuss the question in pairs or in small teams. Students discuss their ideas and what they plan to write. This little prewriting activity ignites ideas for those students who stare blankly at their Journal Writing page. The interpersonal interaction further helps students articulate what they are thinking about the topic and invites students to delve deeper into the topic.

Tell students before they write that they will share their journal entries with a partner or with their team. This motivates many students to improve their entry. Sharing written responses also promotes flexible thinking with open-ended questions, and allows students to hear their peers' responses, ideas and writing styles.

Have students keep a collection of their journal entries in a three-ring binder. This way you can collect them if you wish for assessment or have students go back to reflect on their own learning. If you are using questions across the curriculum, each subject can have its own journal or own section within the binder. Use the provided blackline on the following page for a cover for students' journals or have students design their own.

3. Question Starters

The Question Starters activity page is designed to put the questions in the hands of your students. Use these question starters to scaffold your students' ability to write their own thinking questions. This page includes eight question starters to direct students to generate questions across the levels and types of thinking. This Question Starters activity page can be used in a few different ways:

Individual Questions

Have students independently come up with their own questions. When done, they can trade their questions with a partner. On a separate sheet of paper students answer their partners' questions. After answering, partners can share how they answered each other's questions.

JOURNAL

My Best Thinking

_____
This Journal Belongs to

Higher-Level Thinking Questions for Biology
Kagan Publishing • 1 (800) 933-2667 • www.KaganOnline.com

Pair Questions
Students work in pairs to generate questions to send to another pair. Partners take turns writing each question and also take turns recording each answer. After answering, pairs pair up to share how they answered each other's questions.

Team Questions
Students work in teams to generate questions to send to another team. Teammates take turns writing each question and recording each answer. After answering, teams pair up to share how they answered each other's questions.

Teacher-Led Questions
For young students, lead the whole class in coming up with good higher-level thinking questions.

Teach Your Students About Thinking and Questions
An effective tool to improve students' thinking skills is to teach students about the types of thinking skills and types of questions. Teaching students about the types of thinking skills improves their metacognitive abilities. When students are aware of the types of thinking, they may more effectively plan, monitor, and evaluate their own thinking. When students understand the types of questions and the basics of question construction, they are more likely to create effective higher-level thinking questions. In doing so they develop their own thinking skills and the thinking of classmates as they work to answer each other's questions.

Table of Activities

The Question Cards can be used in a variety of game-like formats to forge students' thinking skills. They can be used for cooperative team and pair work, for whole-class questioning, for independent activities, or at learning centers. On the following pages you will find numerous excellent options to use your Question Cards. As you use the Question Cards in this book, try the different activities listed below to add novelty and variety to the higher-level thinking process.

Team Activities
1. Question Commander 16
2. Fan-N-Pick 18
3. Spin-N-Think 18
4. Three-Step Interview 19
5. Team Discussion 19
6. Think-Pair-Square 20
7. Question-Write-RoundRobin 20

Class Activities
1. Mix-Pair-Discuss 21
2. Think-Pair-Share 21
3. Inside-Outside Circle 22
4. Question & Answer 22
5. Numbered Heads Together 23

Pair Activities
1. RallyRobin 23
2. Pair Discussion 24
3. Question-Write-Share-Discuss 24

Individual Activities
1. Journal Writing 25
2. Independent Answers 25

Learning Centers
1. Question Card Center 26
2. Journal Writing Center 26
3. Question Starters Center 26

Higher-Level Thinking Question Card
Activities

team activity #1

Question Commander

Preferably in teams of four, students shuffle their Question Cards and place them in a stack, questions facing down, so that all teammates can easily reach the Question Cards. Give each team a Question Commander set of instructions (blackline provided on following page) to lead them through each question.

Student One becomes the Question Commander for the first question. The Question Commander reads the question aloud to the team, then asks the teammates to think about the question and how they would answer it. After the think time, the Question Commander selects a teammate to answer the question. The Question Commander can spin a spinner or roll a die to select who will answer. After the teammate gives the answer, Question Commander again calls for think time, this time asking the team to think about the answer. After the think time, the Question Commander leads a team discussion in which any teammember can contribute his or her thoughts or ideas to the question, or give praise or reactions to the answer.

When the discussion is over, Student Two becomes the Question Commander for the next question.

Question Commander
Instruction Cards

Question Commander

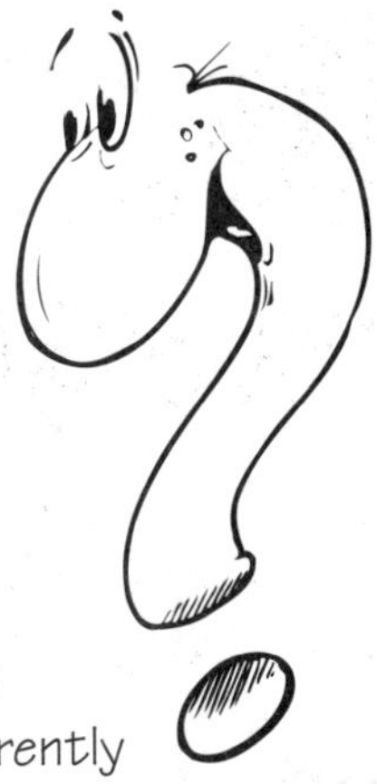

1. **Ask the Question:** Question Commander reads the question to the team.
2. **Think Time:** "Think of your best answer."
3. **Answer the Question:** The Question Commander selects a teammate to answer the question.
4. **Think Time:** "Think about how you would answer differently or add to the answer."
5. **Team Discussion:** As a team, discuss other possible answers or reactions to the answer given.

Question Commander

1. **Ask the Question:** Question Commander reads the question to the team.
2. **Think Time:** "Think of your best answer."
3. **Answer the Question:** The Question Commander selects a teammate to answer the question.
4. **Think Time:** "Think about how you would answer differently or add to the answer."
5. **Team Discussion:** As a team, discuss other possible answers or reactions to the answer given.

Question Commander

1. **Ask the Question:** Question Commander reads the question to the team.
2. **Think Time:** "Think of your best answer."
3. **Answer the Question:** The Question Commander selects a teammate to answer the question.
4. **Think Time:** "Think about how you would answer differently or add to the answer."
5. **Team Discussion:** As a team, discuss other possible answers or reactions to the answer given.

Question Commander

1. **Ask the Question:** Question Commander reads the question to the team.
2. **Think Time:** "Think of your best answer."
3. **Answer the Question:** The Question Commander selects a teammate to answer the question.
4. **Think Time:** "Think about how you would answer differently or add to the answer."
5. **Team Discussion:** As a team, discuss other possible answers or reactions to the answer given.

team activity #2

Fan-N-Pick

In a team of four, Student One fans out the question cards, and says, "Pick a card, any card!" Student Two picks a card and reads the question out loud to teammates. After five seconds of think time, Student Three gives his or her answer. After another five seconds of think time, Student Four paraphrases, praises, or adds to the answer given. Students rotate roles for each new round.

team activity #3

Spin-N-Think

Spin-N-Think spinners are available from Kagan to lead teams through the steps of higher-level thinking. Students spin the Spin-N-Think™ spinner to select a student at each stage of the questioning to: 1) ask the question, 2) answer the question, 3) paraphrase and praise the answer, 4) augment the answer, and 5) discuss the question or answer. The Spin-N-Think™ game makes higher-level thinking more fun, and holds students accountable because they are often called upon, but never know when their number will come up.

team activity #4

Three-Step Interview

After the question is read to the team, students pair up. The first step is an interview in which one student interviews the other about the question. In the second step, students remain with their partner but switch roles: The interviewer becomes the interviewee. In the third step, the pairs come back together and each student in turn presents to the team what their partner shared. Three-Step Interview is strong for individual accountability, active listening, and paraphrasing skills.

team activity #5

Team Discussion

Team Discussion is an easy and informal way of processing the questions: Students read a question and then throw it open for discussion. Team Discussion, however, does not ensure that there is individual accountability or equal participation.

Higher-Level Thinking Questions for Biology
Kagan Publishing • 1 (800) 933-2667 • www.KaganOnline.com

team activity #6

Think-Pair-Square

One student reads a question out loud to teammates. Partners on the same side of the table then pair up to discuss the question and their answers. Then, all four students come together for an open discussion about the question.

team activity #7

Question-Write-RoundRobin

Students take turns asking the team the question. After each question is asked, each student writes his or her ideas on a piece of paper. After students have finished writing, in turn they share their ideas. This format creates strong individual accountability because each student is expected to develop and share an answer for every question.

class activity #1

Mix-Pair-Discuss

Each student gets a different Question Card. For 16 to 32 students, use two sets of questions. In this case, some students may have the same question which is OK. Students get out of their seats and mix around the classroom. They pair up with a partner. One partner reads his or her Question Card and the other answers. Then they switch roles. When done they trade cards and find a new partner. The process is repeated for a predetermined amount of time. The rule is students cannot pair up with the same partner twice. Students may get the same questions twice or more, but each time it is with a new partner. This strategy is a fun, energizing way to ask and answer questions.

class activity #2

Think-Pair-Share

Think-Pair-Share is teacher-directed. The teacher asks the question, then gives students think time. Students then pair up to share their thoughts about the question. After the pair discussion, one student is called on to share with the class what was shared in his or her pair. Think-Pair-Share does not provide as much active participation for students as Think-Pair-Square because only one student is called upon at a time, but is a nice way to do whole-class sharing.

class activity #3

Inside-Outside Circle

Each student gets a Question Card. Half of the students form a circle facing out. The other half forms a circle around the inside circle; each student in the outside circle faces one student in the inside circle. Students in the outside circle ask inside circle students a question. After the inside circle students answer the question, students switch roles questioning and answering. After both have asked and answered a question, they each praise theother's answers and then hold up a hand indicating they are finished. When most students have a hand up, have students trade cards with their partner and rotate to a new partner. To rotate, tell the outside circle to move to the left. This format is a lively and enjoyable way to ask questions and have students listen to the thinking of many classmates.

class activity #4

Question & Answer

This might sound familiar: Instead of giving students the Question Cards, the teacher asks the questions and calls on one student at a time to answer. This traditional format eliminates simultaneous, cooperative interaction, but may be good for introducing younger students to higher-level questions.

22   Higher-Level Thinking Questions for Biology
Kagan Publishing • 1 (800) 933-2667 • www.KaganOnline.com

class activity #5

Numbered Heads Together

Students number off in their teams so that every student has a number. The teacher asks a question. Students put their "heads together" to discuss the question. The teacher then calls on a number and selects a student with that number to share what his or her team discussed.

pair activity #1

RallyRobin

Each pair gets a set of Question Cards. Student A in the pair reads the question out loud to his or her partner. Student B answers. Partners take turns asking and answering each question.

pair activity #2

Pair Discussion

Partners take turns asking the question. The pair then discusses the answer together. Unlike RallyRobin, students discuss the answer. Both students contribute to answering and to discussing each other's ideas.

pair activity #3

Question-Write-Share-Discuss

One partner reads the Question Card out loud to his or her teammate. Both students write down their ideas. Partners take turns sharing what they wrote. Partners discuss how their ideas are similar and different.

individual activity #1

Journal Writing

Students pick one Question Card and make a journal entry or use the question as the prompt for an essay or creative writing. Have students share their writing with a partner or in turn with teammates.

individual activity #2

Independent Answers

Students each get their own set of Questions Cards. Pairs or teams can share a set of questions, or the questions can be written on the board or put on the overhead projector. Students work by themselves to answer the questions on a separate sheet of paper. When done, students can compare their answers with a partner, teammates, or the whole class.

Center Ideas

1. Question Card Center
At one center, have the Question Cards and a Spin-N-Think™ spinner, Question Commander instruction card, or Fan-N-Pick instructions. Students lead themselves through the thinking questions. For individual accountability, have each student record their own answer for each question.

2. Journal Writing Center
At a second center, have a Journal Writing activity page for each student. Students can discuss the question with others at their center, then write their own journal entry. After everyone is done writing, students share what they wrote with other students at their center.

3. Question Starters Center
At a third center, have a Question Starters page. Split the students at the center into two groups. Have both groups create thinking questions using the Question Starters activity page. When the groups are done writing their questions, they trade questions with the other group at their center. When done answering each other's questions, two groups pair up to compare their answers.

Animals

Organisms of the kingdom *Animalia*, distinguished by characteristics such as locomotion, consumers, fixed structure, and limited growth.

higher-level thinking questions

"What is a scientist after all? It is a curious man looking through a keyhole, the keyhole of nature, trying to know what's going on."

— Jacques Cousteau

Animals
Question Cards

Animals

1 If you could be any mammal besides a human, what would you be? Why?

Animals

2 In one sentence how could you describe Kingdom Animalia?

Animals

3 What steps are involved in the development of vertebrate eggs?

Animals

4 If you could create a new animal, what would it look like and what would its characteristics be?

Higher-Level Thinking Questions for Biology
Kagan Publishing • 1 (800) 933-2667 • www.KaganOnline.com

Animals
Question Cards

Animals

5 What is your opinion on keeping animals caged in zoos?

Animals

6 If you could pick one extinct animal to return to our environment, which would you pick? Why?

Animals

7 In your own words, what are the similarities and differences among the hearts of fish, birds, and mammals?

Animals

8 Compare and contrast the structure of echinoderms and arthropods.

Higher-Level Thinking Questions for Biology
Kagan Publishing • 1 (800) 933-2667 • www.KaganOnline.com

Animals
Question Cards

Animals

9 All animals share several key characteristics; identify these characteristics in picture or symbol form.

Animals

10 What characteristics make crustaceans similar to insects?

Animals

11 Describe why it is more efficient for cells to become specialized in a multicellular animal.

Animals

12 Throughout history, *Porifera* have been classified as plants and animals. What characteristics do they share with each kingdom?

Higher-Level Thinking Questions for Biology
Kagan Publishing • 1 (800) 933-2667 • www.KaganOnline.com

Animals
Question Cards

Animals

13 How is the movement of a squid similar to a balloon releasing gas?

Animals

14 After heavy rain, earthworms are often seen on top of the ground. Why do you think this is so?

Animals

15 If you were the first vertebrate to inhabit the land, what would you have seen?

Animals

16 Compare and contrast the bone structure in a bird and mammal. What benefits do each have for adaptation and survival?

Animals

Journal Writing Question

Write your response to the question below.
Be ready to share your response.

Most of the damage done to coral reefs is caused by humans. Brainstorm a list of strategies that can be implemented to protect reefs. How might these changes impact your own life?

Animals
Question Starters

Use the question starters below to create complete questions.
Send your questions to a partner or to another team to answer.

1. Do you think birds will

2. Why are fish

3. How do mammals

4. What animals may

5. What do you wonder about the phyla

6. How is   like

7. What is the relationship between

8. What might happen if

Biochemistry

The chemistry that deals with the chemical compounds and reactions that relate to living things.

higher-level thinking questions

> **Satisfaction of one's curiosity is one of the greatest sources of happiness in Life.**
>
> — Linus Pauling

Biochemistry Question Cards

Biochemistry

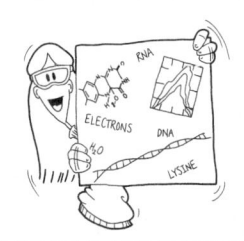

1. If you could only eat one type of carbohydrate (monosaccharides, disaccharides, polysaccharides) for one week, what kind would you choose? Why?

Biochemistry

2. Do you think it is possible for humans to survive on only simple sugars?

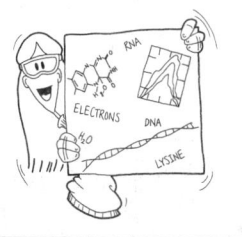

Biochemistry

3. What do you think would happen if all lipids were to disappear from Earth?

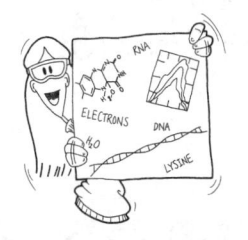

Biochemistry

4. In what ways are all four classes of macromolecules alike?

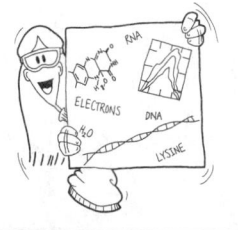

Higher-Level Thinking Questions for Biology
Kagan Publishing • 1 (800) 933-2667 • www.KaganOnline.com

Biochemistry Question Cards

Biochemistry

5 What makes the nucleic acids so different from the other types of macromolecules?

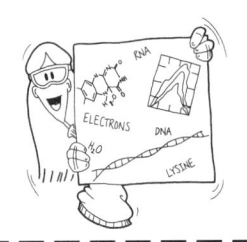

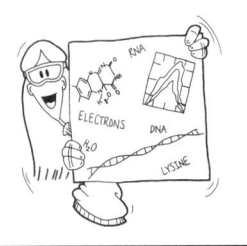

Biochemistry

6 What are some of the possible reasons why carbon is the backbone for all of the macromolecules?

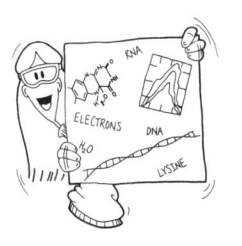

Biochemistry

7 In what ways is a chemical buffer similar to a living cell when adjusting to pH changes in the environment?

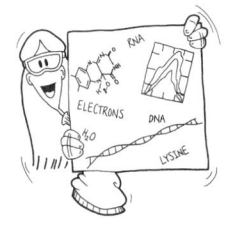

Biochemistry

8 Imagine you are an amino acid in a polypeptide chain. What would you experience as your chain begins to fold to form the 3-D structure of a complex protein?

Biochemistry
Question Cards

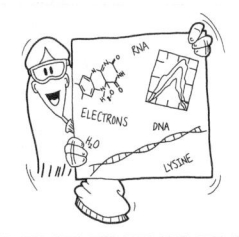

Biochemistry

9 What similarities and differences do RNA and DNA share in terms of structure and function?

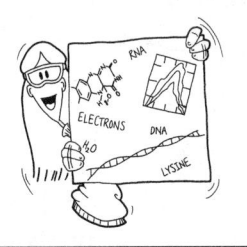

Biochemistry

10 Design an experiment that illustrates why is it more difficult for a starch to move across a cell membrane than a monosacaccharide.

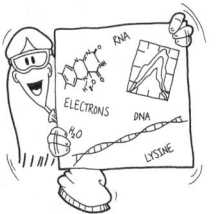

Biochemistry

11 How has the waterproof nature of lipids impacted your daily life?

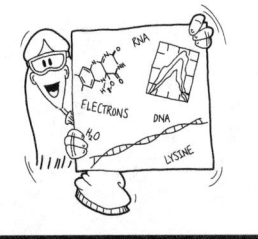

Biochemistry

12 What would life be like without proteins?

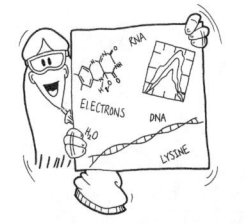

Higher-Level Thinking Questions for Biology
Kagan Publishing • 1 (800) 933-2667 • www.KaganOnline.com

39

Biochemistry Question Cards

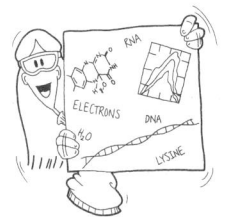

Biochemistry

13 What do you predict would happen to the structure and function of a protein if it were exposed to an extremely acidic pH? Explain.

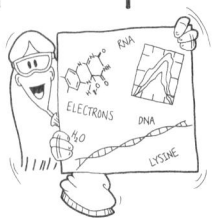

Biochemistry

14 What would it feel like to be a cell membrane (made up of phospholipids) that was plunged into an extremely cold environment?

Biochemistry

15 What might the world be like if all organisms did not share the same basic genetic component (DNA)?

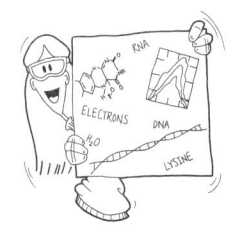

Biochemistry

16 If you were stranded on a lifeboat, which would you prefer to have to eat: a pound of lard or a pound of pasta? Why?

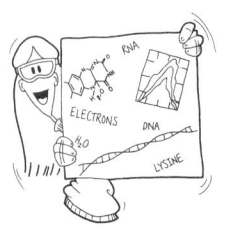

Biochemistry
Journal Writing Question

Write your response to the question below.
Be ready to share your response.

Describe what the world would be like if ice did not float.

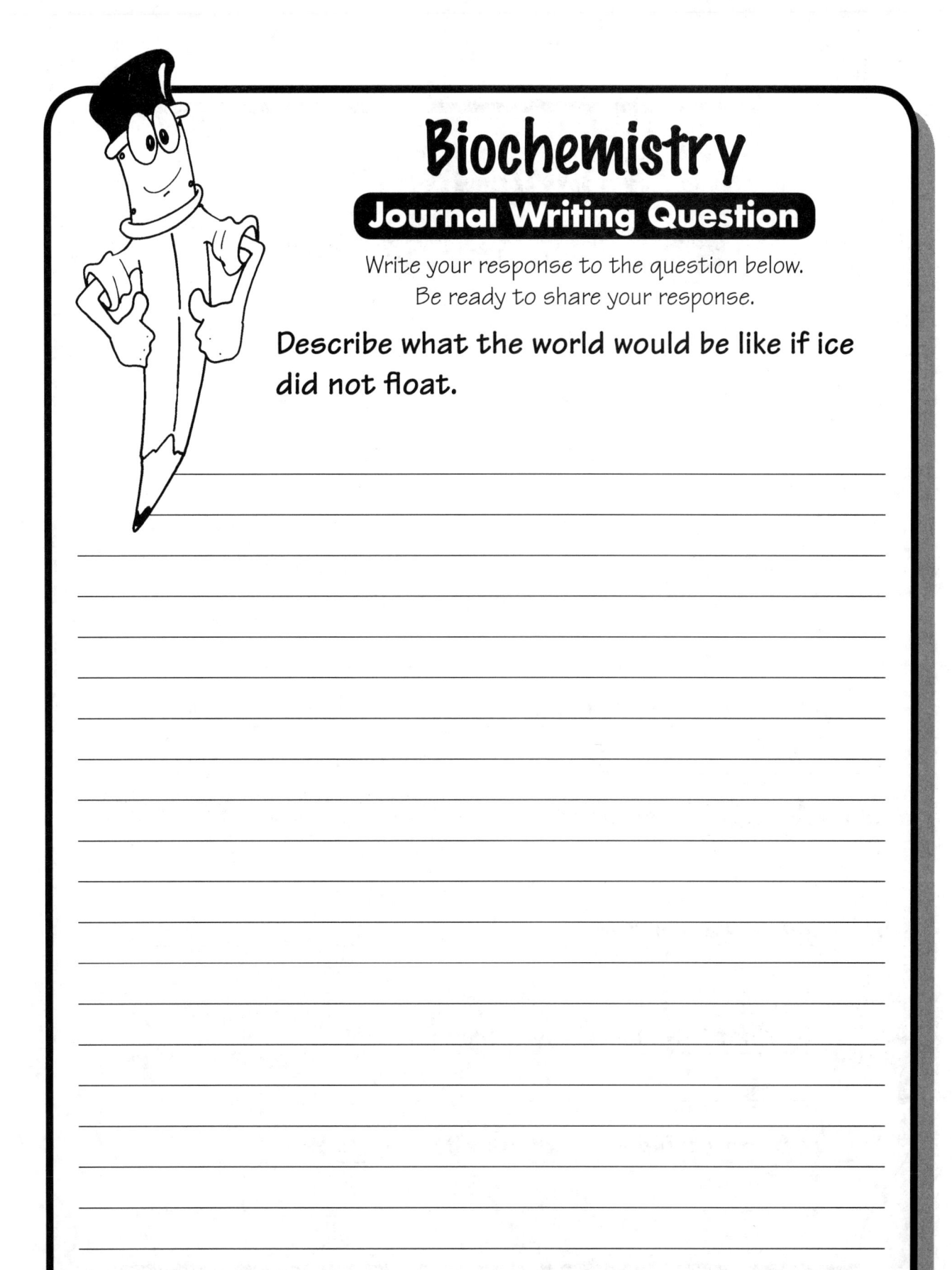

Biochemistry
Question Starters

Use the question starters below to create complete questions.
Send your questions to a partner or to another team to answer.

1. Why do proteins

2. Should DNA

3. What do you think would happen if hydrolysis

4. If you were an apple filled with fructose

5. What practical applications do macromolecules

6. How could you improve

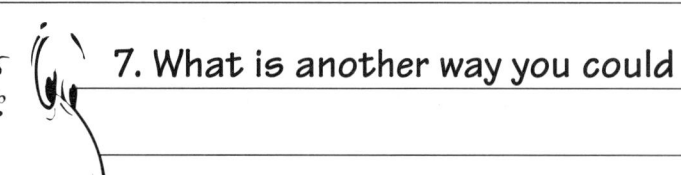

7. What is another way you could

8. How could you express with a symbol

Bioenergetics

The study of how organisms carry out energy transformations.

higher-level thinking questions

> **If everybody is thinking alike, then somebody isn't thinking.**
>
> — George S. Patton, General

Bioenergetics Question Cards

Bioenergetics

1 Do you think it is possible for scientists to incorporate chloroplasts into the skin cells of humans? What would this mean for humans?

Bioenergetics

2 Why are C4 plants so successful in the summertime in the Temperate Regions?

Bioenergetics

3 What adaptations make CAM plants so successful in the desert?

Bioenergetics

4 Why is it theorized that chloroplasts may be descendents of photosynthetic bacteria?

Higher-Level Thinking Questions for Biology
Kagan Publishing • 1 (800) 933-2667 • www.KaganOnline.com

Bioenergetics Question Cards

Bioenergetics

5 What types of plants would be most beneficial for use in a space station? What would they be used for?

Bioenergetics

6 What is the significance in the layering of cell types in a leaf?

Bioenergetics

7 As a molecule of water, explain your journey through a plant from the moment of entry in the root to your involvement in photosynthesis.

Bioenergetics

8 When experiencing oxygen debt, why do human cells not carry out the process of alcohol fermentation?

Bioenergetics Question Cards

Bioenergetics

9 What role does the mitochondria play in cellular respiration?

Bioenergetics

10 If you were a glucose molecule, how would your structure change as you moved from the cytoplasm to the mitochondria?

Bioenergetics

11 Pretend you are a CO_2 molecule. What would your journey through a leaf look like?

Bioenergetics

12 How does your knowledge of chemical metabolism relate to your life?

Bioenergetics Question Cards

Bioenergetics

13 Why is aerobic respiration so much more efficient than anaerobic respiration?

Bioenergetics

14 What are the pluses and minuses of alcoholic and lactic acid fermentation?

Bioenergetics

15 Why does the structure of ATP allow it to be the "currency" of the cell?

Bioenergetics

16 If you were a reporter, how would you describe the Kreb's Cycle as it is happening?

Bioenergetics
Journal Writing Question

Write your response to the question below.
Be ready to share your response.

Imagine that you are a plant. Describe the process of producing glucose (photosynthesis) and then generating ATP for cell use (cellular respiration) as you experience these processes.

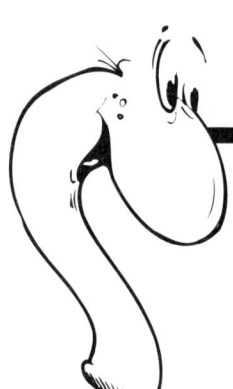

Bioenergetics
Question Starters

Use the question starters below to create complete questions. Send your questions to a partner or to another team to answer.

1. If animals could photosynthesize

2. How does photosynthesis affect

3. What would happen if glycolysis were

4. How is cellular respiration like

5. What patterns do you see in

6. What connections can you make between

7. In your own words, what is

8. In one word, how would you describe

Biotechnology and Ethics

The application of the principles of engineering and technology to the life sciences and the implications of ethical decision-making.

higher-level thinking questions

> **If we all did the things we are capable of, we would astound ourselves.**
>
> — Thomas Edison

Biotechnology and Ethics
Question Cards

Biotechnology and Ethics

1 What are possible advantages and disadvantages to stem cell research?

Biotechnology and Ethics

2 If you could create a genetically modified organism, what would you create? How would this organism impact the environment?

Biotechnology and Ethics

3 You are a government official. What would you say to the governing body to help pass a law that funds genetic engineering research facilities?

Biotechnology and Ethics

4 Should insurance companies have the right to genetically screen future clients? Explain your answer.

Higher-Level Thinking Questions for Biology
Kagan Publishing • 1 (800) 933-2667 • www.KaganOnline.com

53

Biotechnology and Ethics
Question Cards

Biotechnology and Ethics

5 What impact might DNA manipulation have on the ecosystem if it were used to re-create extinct organisms?

Biotechnology and Ethics

6 If you could shrink yourself small enough to jump inside of a DNA molecule, what would you see?

Biotechnology and Ethics

7 Do you think it is morally acceptable to clone human cells or organs inside of another organism for medical use?

Biotechnology and Ethics

8 If you could obtain Human Growth Hormone (HGH) from an anonymous source for athletic improvement, would you use it? Why or why not?

Biotechnology and Ethics
Question Cards

Biotechnology and Ethics

9 If you were the head of research for a cosmetic company and animal testing would lead to an increase in product sales, would you test your products on animals? Explain.

Biotechnology and Ethics

10 Would you have your (or your spouse's) sperm sorted so that you could choose to have a boy or girl? Explain why or why not.

Biotechnology and Ethics

11 Would you choose to create a "designer baby" using biotechnology (to either prevent a genetic disorder or to improve your child's genes)? Explain.

Biotechnology and Ethics

12 Do the benefits of having a pest-resistant crop outweigh the destruction and possible extinction of another species? Explain your answer.

Higher-Level Thinking Questions for Biology
Kagan Publishing • 1 (800) 933-2667 • www.KaganOnline.com

55

Biotechnology and Ethics
Question Cards

Biotechnology and Ethics

13 Are the needs of humans more important than those of the other species found living in the forest? Explain your answer.

Biotechnology and Ethics

14 Should humans be able to genetically alter an organism's physical appearance for profit?

Biotechnology and Ethics

15 What measures do you think should be taken to control a population's growth?

Biotechnology and Ethics

16 Do you think it is right to pay humans or use "death row" inmates to become "guinea pigs" for drug experimentation and research testing? Explain.

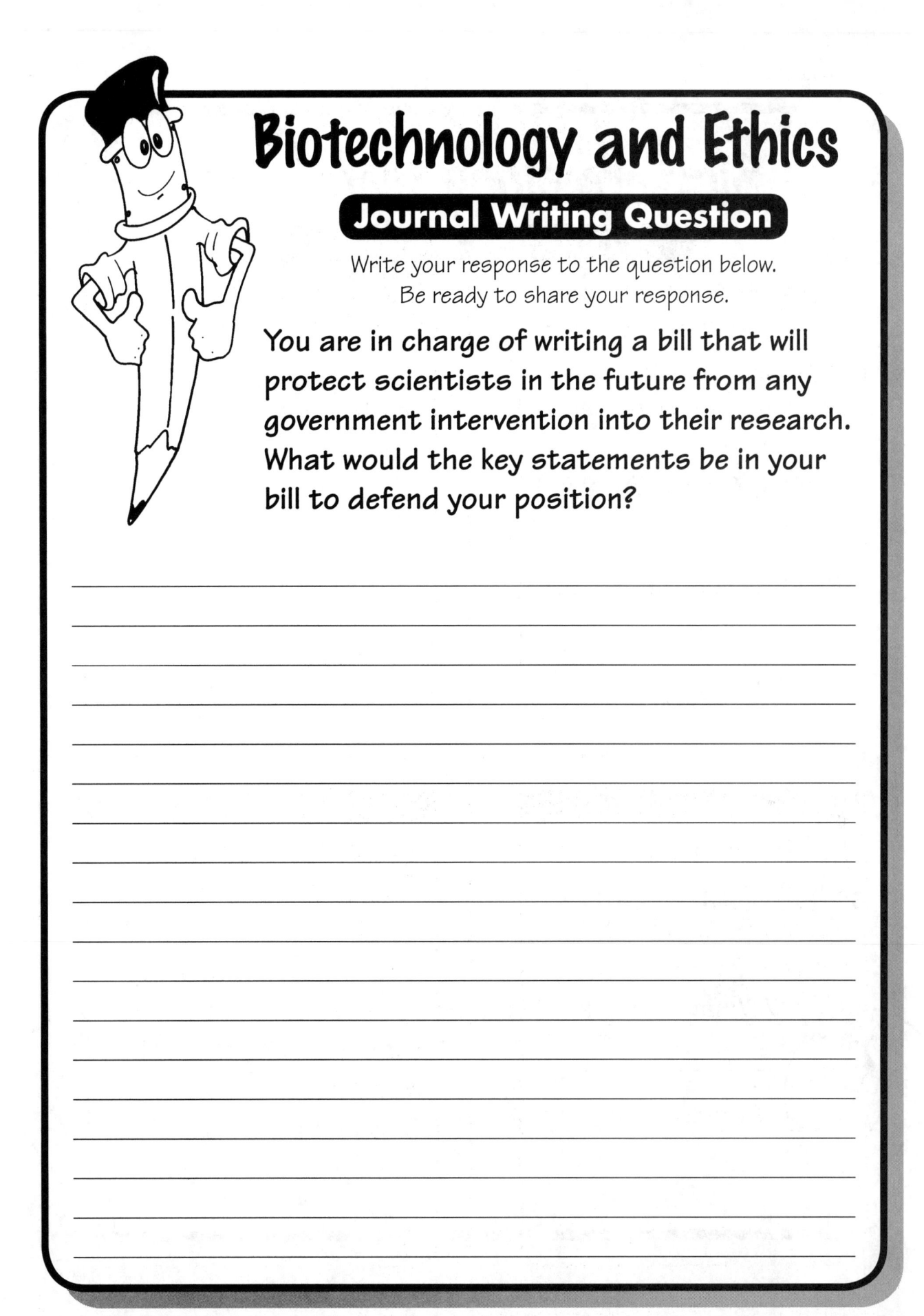

Biotechnology and Ethics

Journal Writing Question

Write your response to the question below.
Be ready to share your response.

You are in charge of writing a bill that will protect scientists in the future from any government intervention into their research. What would the key statements be in your bill to defend your position?

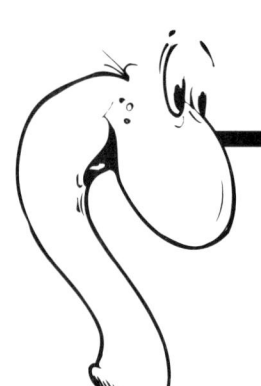

Biotechnology and Ethics
Question Starters

Use the question starters below to create complete questions.
Send your questions to a partner or to another team to answer.

1. If you were a geneticist

2. What role does DNA play in

3. What do you wonder about

4. What dangers

5. What would be different if genetic information were

6. What would you do if

7. What might happen if cancer

8. How would you feel if

Body Systems

A group of organs that work together to perform one type of job for an organism.

higher-level thinking questions

"In questions of science, the authority of a thousand is not worth that humble reasoning of a single individual."

— Galileo Galilei

Body Systems
Question Cards

Body Systems

1 How is the human body like an automobile? How is it different?

Body Systems

2 The nervous system is the body's most important system. Do you agree or disagree? Why?

Body Systems

3 If you could change one thing about the digestive system, what would you change? Why?

Body Systems

4 If you could change one thing about the circulatory system, what would you change? Why?

Higher-Level Thinking Questions for Biology
Kagan Publishing • 1 (800) 933-2667 • www.KaganOnline.com

Body Systems Question Cards

Body Systems

5 Complete the following comparison: The respiratory system is like _____.

Body Systems

6 In your own words, explain the reflex arc.

Body Systems

7 What might be the implications on humans if pain were not as extreme?

Body Systems

8 If you had asthma, how would it impact your daily life?

Body Systems
Question Cards

Body Systems

9 How is your skin similar to a cell membrane? How is it different?

Body Systems

10 You are a T-cell fighting off an infection inside a body. What is your battle like? Describe it.

Body Systems

11 Using your body, illustrate the movement of a hinge, ball and socket, pivot, and saddle joint. How are the movements different?

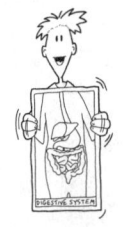

Body Systems

12 Pretend you are a nerve cell that has experienced the effects of a depressant. Describe what has happened to your function.

Body Systems
Question Cards

Body Systems

13 You are a red blood cell traveling through the circulatory system. How would you describe your journey?

Body Systems

14 How does structure follow function in human body systems?

Body Systems

15 What would it feel like to be in the shoes of someone who did not have the sense of _____ (hearing, sight, taste, touch, smell)?

Body Systems

16 What is the relationship between the skeletal and muscular systems in completing a movement of the body?

Body Systems
Journal Writing Question

Write your response to the question below. Be ready to share your response.

If you were a hamburger being consumed by the body, describe what your journey through the digestive system would be like.

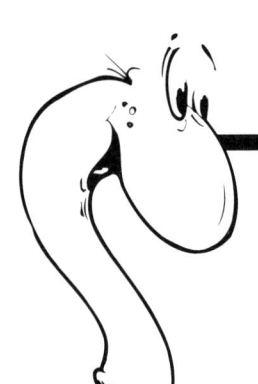

Body Systems
Question Starters

Use the question starters below to create complete questions.
Send your questions to a partner or to another team to answer.

1. If you ate a potato chip

2. How does a neuron relate to

3. Why is hydrochloric acid

4. If red blood cells were

5. What would it feel like to be

6. How does air pollution affect

7. In your own words, what is

8. What questions do you have about

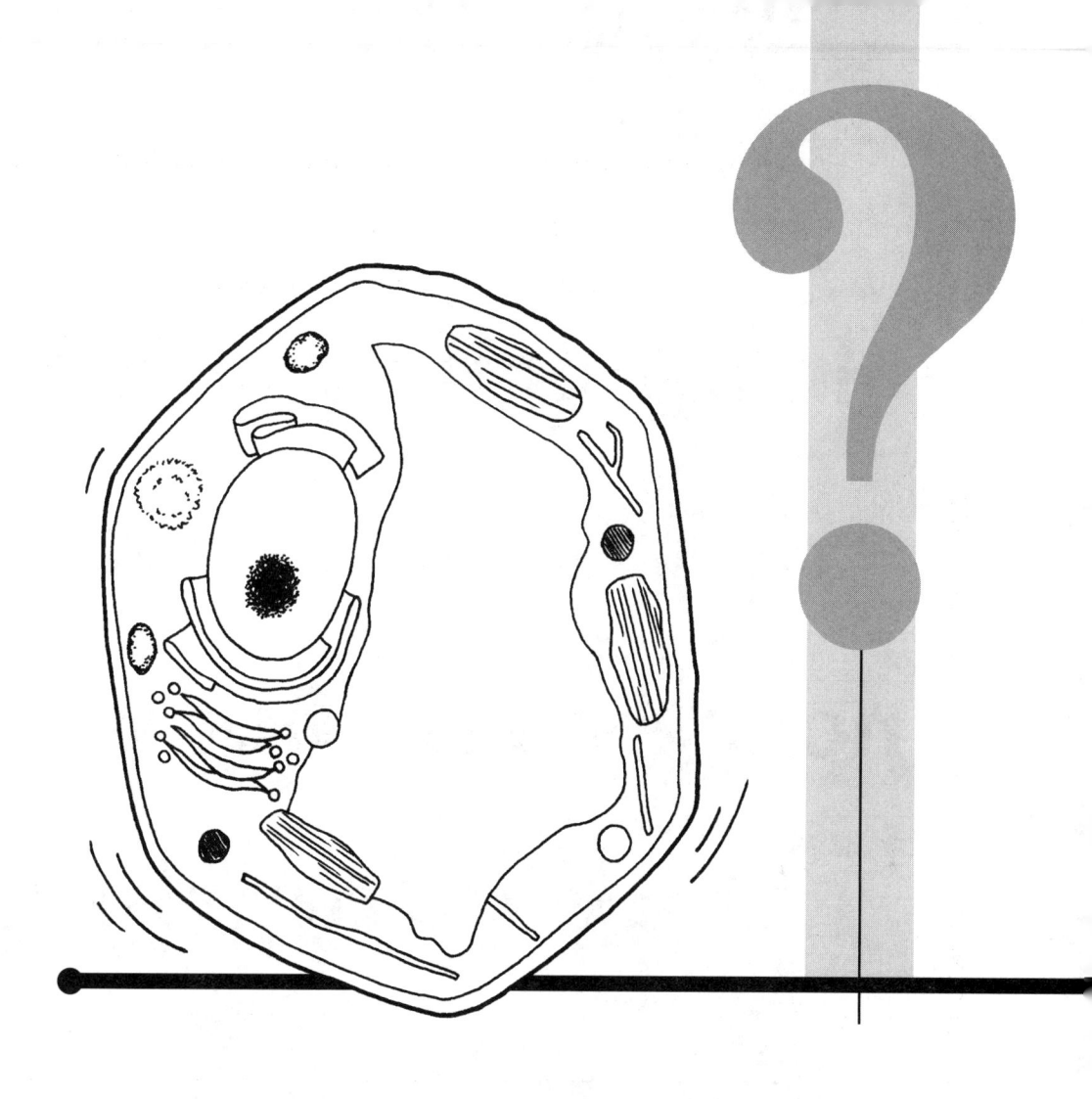

Cells

The smallest functional units of life.

higher-level thinking questions

> **Imagination will often carry us to worlds that never were. But without it we go nowhere.**
>
> — Carl Sagan

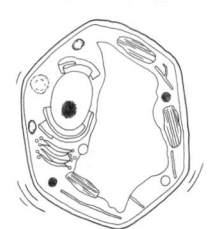

Cells
Question Cards

Cells

1. If you were a chromosome, what would the inside of your cell look like as your cell proceeded through the steps of cellular division?

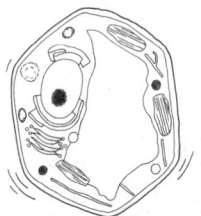

Cells

2. Pretend that you are a molecule of water. Act out with your body what would happen to this molecule as it moves across the cell membrane.

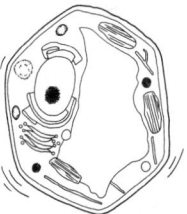

Cells

3. How large do you think a single cell could be? How small? Defend your answer.

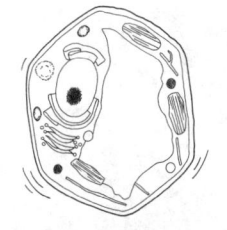

Cells

4. Compare and contrast the life cycles of cells and humans.

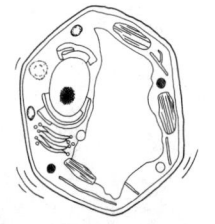

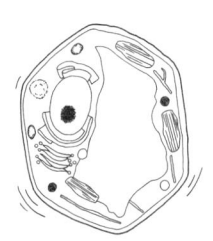

Cells
Question Cards

Cells

5 Pretend you are a molecule of RNA. Describe your journey from the nucleus to the ribosome.

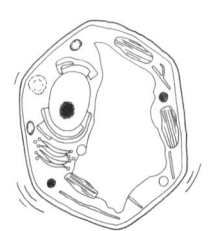

Cells

6 Why is it so important that cells group together to form tissues?

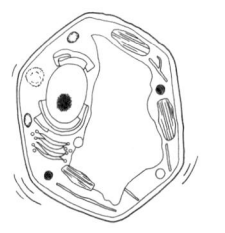

Cells

7 What might be the result of removing mitochondria from the cells of an animal?

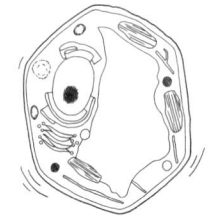

Cells

8 If you could create a new function that a cell could perform, what would that function be? Describe.

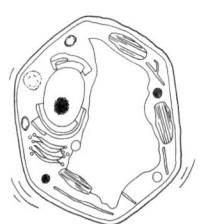

Cells
Question Cards

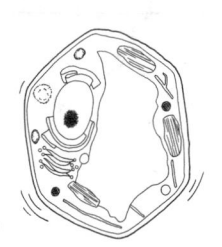

Cells

9 What would it be like if you were a muscle cell?

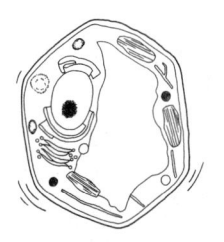

Cells

10 How is a cell like a small town? How is it different?

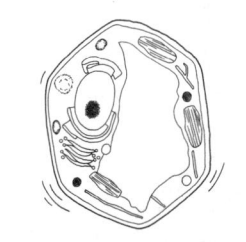

Cells

11 What experiment could you design to test movement across a cell membrane?

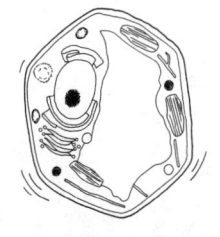

Cells

12 What would it feel like to be the nucleus of a cell that has been taken over by cancer?

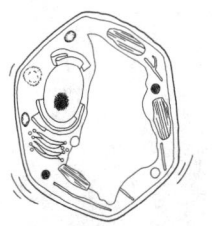

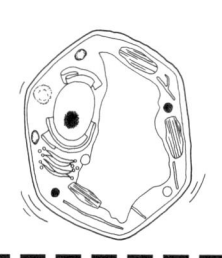

Cells
Question Cards

Cells

13 What are the advantages and disadvantages of animal cells only having a cell membrane (compared to a cell wall found in plants)?

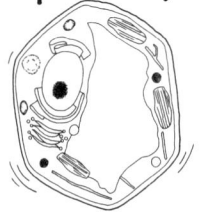

Cells

14 What would happen if all cells had a *single* phospholipid layer making up their membrane?

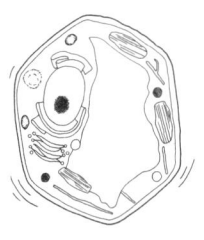

Cells

15 In one sentence, how would you describe the cell theory? Why is this theory called the foundation of modern biology?

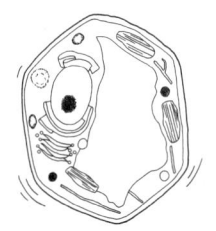

Cells

16 In what ways are a tile mosaic and a plasma membrane similar to each other?

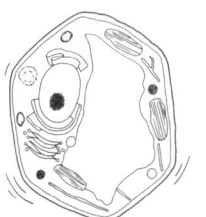

Cells

Journal Writing Question

Write your response to the question below.
Be ready to share your response.

Predict what might happen to a plasma membrane if the phospholipid's tails faced toward the outside of the cell. Explain your thoughts.

Cells
Question Starters

Use the question starters below to create complete questions. Send your questions to a partner or to another team to answer.

1. Do you think cells

2. Should the cell membrane

3. What would happen if the nucleus

4. Why do organelles

5. What patterns can you find

6. How would you describe the organization of

7. What impact might placing a cell in

8. What role

Classification

The science of grouping and categorizing life based on similar characteristics.

higher-level thinking questions

> **Anyone who has never made a mistake has never tried anything new.**
>
> — Albert Einstein

Classification
Question Cards

Classification

1 If you discovered a new organism, what criteria would you use to classify it?

Classification

2 If you had to classify the lichen as an individual organism, what one kingdom would you put it in? Why?

Classification

3 If you had to design a brand new classification system for clothing in a department store, how would you do it? Why?

Classification

4 If life were found on Mars, what criteria do you think should be used to classify it?

Classification
Question Cards

Classification

5 Do you think that viruses should be included or excluded from our classification system? Why or why not?

Classification

6 What is your opinion about using Latin names for modern-day classification? Is it good or bad, helpful or hurtful? Explain.

Classification

7 If you were a zoologist, what features of animals would you use to classify them?

Classification

8 Do you think it could be possible for organisms to be classified in more than one kingdom? Why or why not?

Classification Question Cards

Classification

9 If you could change one thing about the modern classification system, what would it be? Why?

Classification

10 What are some of the possible reasons for using the Binomial System of Nomenclature?

Classification

11 If you were a plant breeder and created a new species or subspecies of a flower, what would you name it? What would be the reason for your choice?

Classification

12 It is very important for humans to classify living organisms. Do you agree or disagree? Why?

Classification Question Cards

Classification

13 How can you apply classification systems in your everyday life?

Classification

14 If you were hired to redesign the classification system used in a chain of music stores, how would your system work?

Classification

15 In what ways do you think a classification system is similar to the system used in a supermarket? How is it different?

Classification

16 Create an analogy for classification: "Classification is to _____ as _____ is to _____."

# Classification
Journal Writing Question

Write your response to the question below.
Be ready to share your response.

What changes do you predict would take place in our current system of classification if the DNA sequence were known for every living species?

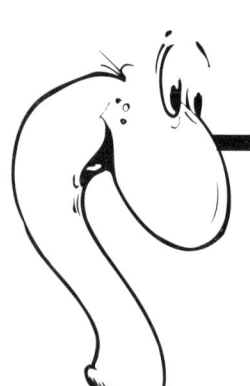

Classification
Question Starters

Use the question starters below to create complete questions.
Send your questions to a partner or to another team to answer.

1. Should the classification system

2. What would be the impact on classification if

3. Do you think it is possible for Latin names to be

4. How is the classification of mammals like

5. What impact might

6. If you wanted to know about

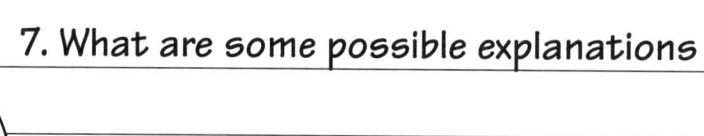

7. What are some possible explanations

8. In what order would you rank

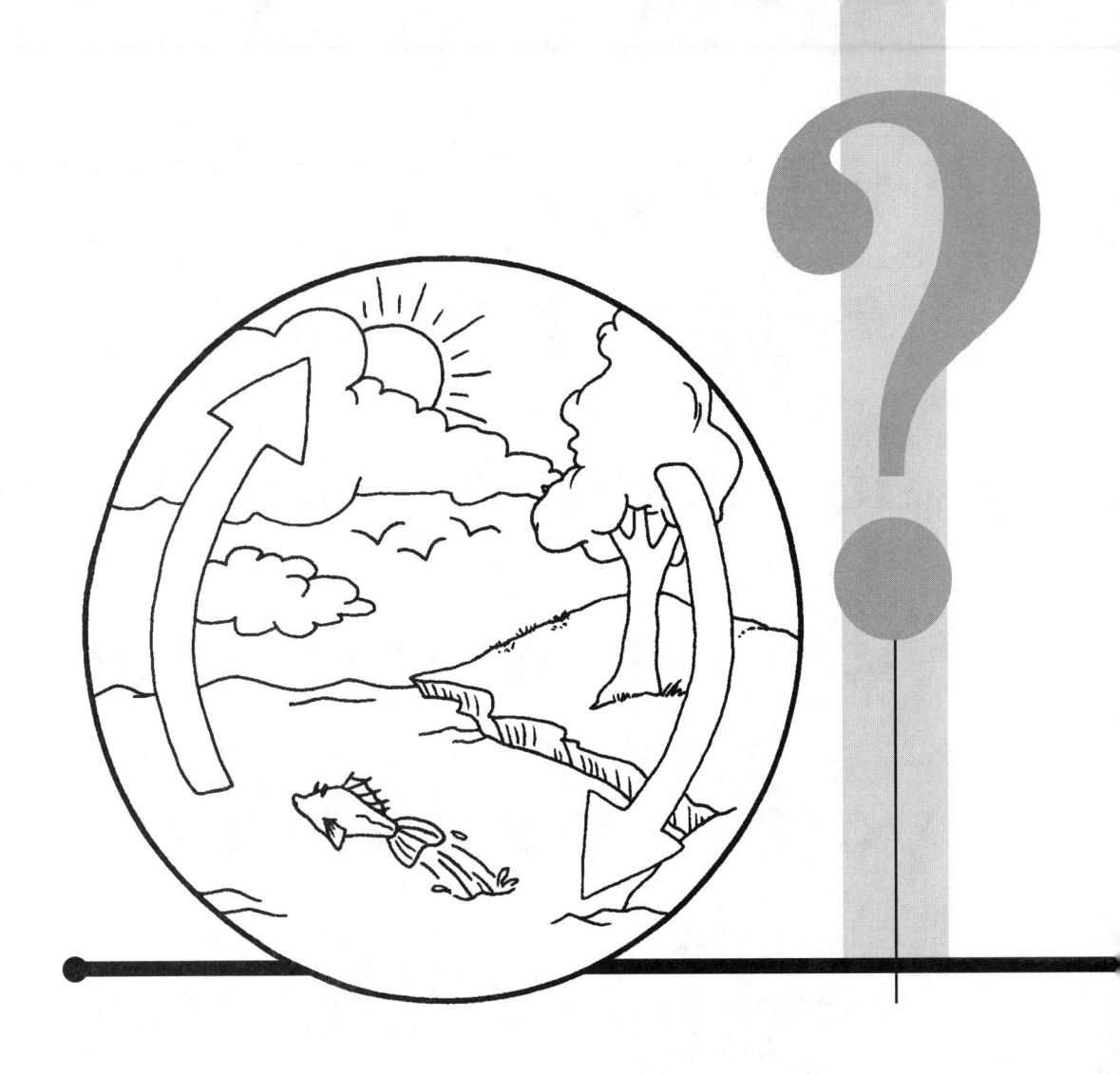

Ecology

The scientific study of interactions between organisms and their environment.

higher-level thinking questions

"Treat the Earth well: it was not given to you by your parents, it was loaned to you by your children. We do not inherit the Earth from our Ancestors, we borrow it from our Children."

— Ancient Native American Proverb

Ecology
Question Cards

Ecology

1 Describe what you would see around you if you were a producer in a community undergoing primary succession.

Ecology

2 Is global warming a myth or is it real? Defend your position.

Ecology

3 If you could create your own biome, what would it be like? Describe the abiotic and biotic factors there.

Ecology

4 What do you think would happen to the water cycle and living organisms if evaporation happened at a lower than normal temperature?

Ecology
Question Cards

Ecology

5 What is your role in the food web?

Ecology

6 If you were a wolf, how would a deadly viral outbreak in the rabbit population affect you?

Ecology

7 Why is it so important to have producers, consumers, and decomposers in an ecosystem?

Ecology

8 Do you believe that humans are a global community? Explain.

Ecology
Question Cards

Ecology

9 Ecologically speaking, what do you think the phrase "Think like a mountain" might mean?

Ecology

10 How would your perspective of the environment change if you were only one inch tall? Explain.

Ecology

11 What would an insect in the rainforest have to say to us if it could speak?

Ecology

12 What ecological impact could a hurricane have on a stable ecosystem?

Higher-Level Thinking Questions for Biology
Kagan Publishing • 1 (800) 933-2667 • www.KaganOnline.com

Ecology
Question Cards

Ecology

13 How do human values impact ecological decisions for our global community?

Ecology

14 As a government official, what would be your top-ten list of ecological priorities?

Ecology

15 From an ecological point of view, explain how diversity leads to stability.

Ecology

16 How could you express in a drawing the importance of conservation on the ecological health of planet Earth?

Ecology

Journal Writing Question

Write your response to the question below.
Be ready to share your response.

You are part of a task force that will decide what to do with a parcel of public land. A portion of the community wants to preserve the area as a wildlife refuge while another group wants a town center to be built. What ecological considerations should be made during the decision-making process?

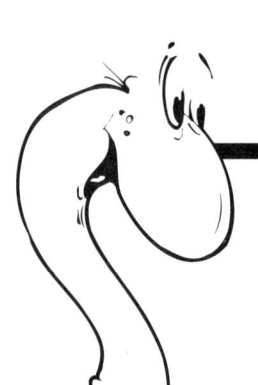

Ecology
Question Starters

Use the question starters below to create complete questions.
Send your questions to a partner or to another team to answer.

1. What if the desert

2. Should predators

3. How can the carbon cycle

4. Why do food chains

5. If you lived in the rainforest

6. If you were the size of a

7. If the sun vanished

8. How is   different from

Evolution

The continuous genetic adaptations of organisms over time in response to changes in the environment.

higher-level thinking questions

> **To think is to differ.**
>
> — Clarence Darrow

Evolution Question Cards

Evolution

1 Do you think that teaching evolution should be allowed in a public school? Explain your opinion.

Evolution

2 You are a female frog that has inherited a genetic mutation that has caused you to ignore the mating calls of male frogs. How will this impact the future of your species?

Evolution

3 How does natural selection cause a population to adapt to its environment over many generations?

Evolution

4 Is evolution a scientific fact or is it just a theory? Explain.

Evolution
Question Cards

Evolution

5. If you were an eyewitness at the time of the Big Bang, what would you have seen?

Evolution

6. What do you think it would be like if Neanderthals were still alive today?

Evolution

7. Why do you think it is theorized that prokaryotes were the first forms of life?

Evolution

8. If you could be transported back in time, what one part of the evolution of life would you most want to witness? Why?

Higher-Level Thinking Questions for Biology
Kagan Publishing • 1 (800) 933-2667 • www.KaganOnline.com

Evolution
Question Cards

Evolution

9 If life exists in other places in the universe, do you believe life evolves like it does here on Earth or would it evolve differently? If differently, how so?

Evolution

10 Humans are evolving. How will humans be different in 100 years from now? 1,000 years?

Evolution

11 Of all the factors that can change a population's gene pool, which one do you think is most significant in terms of human evolution? Why? What about bacteria?

Evolution

12 If you could create an adaptation for any organism, what would it be? Why?

Higher-Level Thinking Questions for Biology
Kagan Publishing • 1 (800) 933-2667 • www.KaganOnline.com

Evolution
Question Cards

Evolution

13 More and more children are being diagnosed with asthma. How could evolution play a role? What possible evolutionary benefits could exist?

Evolution

14 A nonnative catfish (known to be an aggressive predator) was caught in a local river. Should scientists work to capture the species or is this natural selection at work?

Evolution

15 How could you explain why some flowers look like the insects that are their primary pollinators?

Evolution

16 In your opinion, what would be a better evolutionary strategy: being well-adapted to a single environment and food source, or having very diverse adaptations? Explain.

Evolution

Journal Writing Question

Write your response to the question below.
Be ready to share your response.

Pretend you have just traveled back in time to the year 1835. You are walking around on the Galapagos Islands and bump into Charles Darwin. What would you tell him about biology? Why?

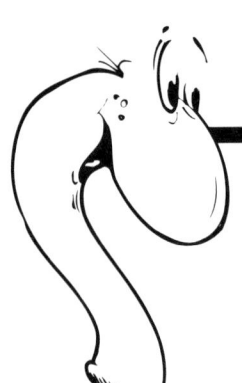

Evolution
Question Starters

Use the question starters below to create complete questions. Send your questions to a partner or to another team to answer.

1. If you were Charles Darwin

2. Do you think it is possible for primates to

3. What do you think would happen if bacteria

4. How has your thinking changed about

5. What is your opinion about

6. Evolution is like

7. What connections can you make between

8. In your life, how would you apply

Fungi

The kingdom of life composed of heterotrophs; many obtain nutrients from dead organic matter.

higher-level thinking questions

> **Judge a man by his questions rather than his answers.**
>
> — Voltaire

Fungi
Question Cards

Fungi

1 What characteristics do all fungi share that put them in the same kingdom?

Fungi

2 What is the main purpose of a mushroom?

Fungi

3 In one sentence, how could you describe mycelium?

Fungi

4 In your own words, how could you describe the life cycle of any fungi?

Fungi
Question Cards

Fungi

5 Fungi cause disease such as ringworm and athlete's foot in humans. Fungi cause more harm than they do good. Do you agree or disagree? Why?

Fungi

6 Why is it so important that yeast readily give off carbon dioxide and who is it most important to?

Fungi

7 Where could you have come in contact with fungi yesterday? Explain.

Fungi

8 Ecologically speaking, predict what the world would be like if it were treated with a global fungicide.

Fungi
Question Cards

Fungi

9 In what ways are the kingdoms Fungi and Monera similar?

Fungi

10 How might the ability to produce spores asexually and sexually help the survival of the Fungi kingdom?

Fungi

11 What are the pros and cons of eradicating from the planet any fungus that causes human disease?

Fungi

12 How could you design an experiment to test for the effectiveness of a new fungicide?

Fungi
Question Cards

Fungi

13 In your life, how could you apply your knowledge about fungal life cycles to prevent "catching" any fungal infections?

Fungi

14 Would a fungus-resistant, genetically-modified crop be an advantage or disadvantage for the environment?

Fungi

15 Design a hypothesis as to why fungi and plants have symbiotic relationships.

Fungi

16 Why do you think fungi are no longer classified as a part of the Plant kingdom?

Fungi

Journal Writing Question

Write your response to the question below.
Be ready to share your response.

If you were the head of a pharmaceutical corporation, why would fungi be so important to you?

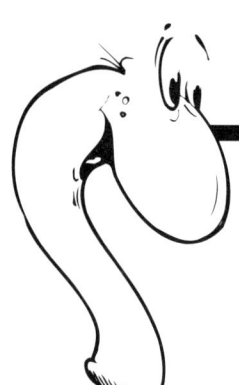

Fungi
Question Starters

Use the question starters below to create complete questions. Send your questions to a partner or to another team to answer.

1. Can all mushrooms

2. How can you explain why fungi

3. Why do the decomposers

4. Do you think fungi

5. In your opinion

6. Why are fungi

7. Fungi are like   because

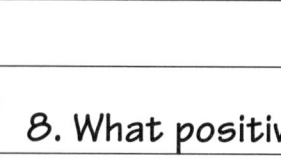

8. What positive impact can yeast

Genetics

The biological study of heredity and variation of organismal characteristics.

higher-level thinking questions

> "You miss 100% of the shots you never take."
>
> — Wayne Gretzky

Genetics
Question Cards

Genetics

1 If you knew that the gene controlling a genetic disorder was present in your family, would you want to undergo genetic testing to determine whether or not you carried the gene? Why or why not?

Genetics

2 What social and ethical issues might new knowledge from genetic experiments raise?

Genetics

3 List all of the genotypes possible from a cross between two pea plants: a hybrid tall and a homozygous recessive short. What would the phenotype ratios be?

Genetics

4 Which of Mendel's four principles do you think is most important to biologists today? Explain.

Higher-Level Thinking Questions for Biology
Kagan Publishing • 1 (800) 933-2667 • www.KaganOnline.com

Genetics
Question Cards

Genetics

5 Drosophila melangostar (fruit fly) has been used in genetic experiments for many years. Should scientists use humans in genetic studies?

Genetics

6 The X chromosome is much larger than the Y chromosome and contains much more genetic information. Why do you think that so much of this information is carried on the X and not the Y chromosome?

Genetics

7 Do you think that a mutant organism would be more or less adapted to survive in its environment? Explain.

Genetics

8 Do you agree or disagree with the farming of genetically modified crops? Explain your answer.

Genetics
Question Cards

Genetics

9 If you were Gregor Mendel, and had by chance chosen to study two traits in peas that were linked, how would this have changed the history of genetic studies?

Genetics

10 Compare and contrast mitosis and meiosis.

Genetics

11 Do you think it would be possible for a person or plant to live with only half of the full chromosome set? Explain.

Genetics

12 If you could crossbreed any two species, which two would you choose to crossbreed? Why?

Genetics
Question Cards

Genetics

13 What would happen in the world if humans could determine which genes were passed on to their children?

Genetics

14 Do you think that the government should control the notification of the public regarding the sale of genetically modified organisms? Explain.

Genetics

15 If you could change any of your genes, which would you change? Why?

Genetics

16 What are some of the benefits and drawbacks that the creation of a DNA identity bank would create?

Genetics

Journal Writing Question

Write your response to the question below.
Be ready to share your response.

Describe what it would be like working as a geneticist without the knowledge of DNA.

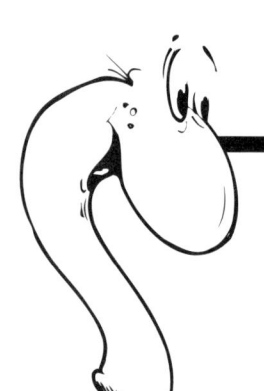

Genetics
Question Starters

Use the question starters below to create complete questions.
Send your questions to a partner or to another team to answer.

1. What would you do if cloning

2. If you were a geneticist

3. What do chromosomes

4. How could a gene

5. What would happen if

6. Do you think that mutations

7. Why is crossbreeding

8. What impact might

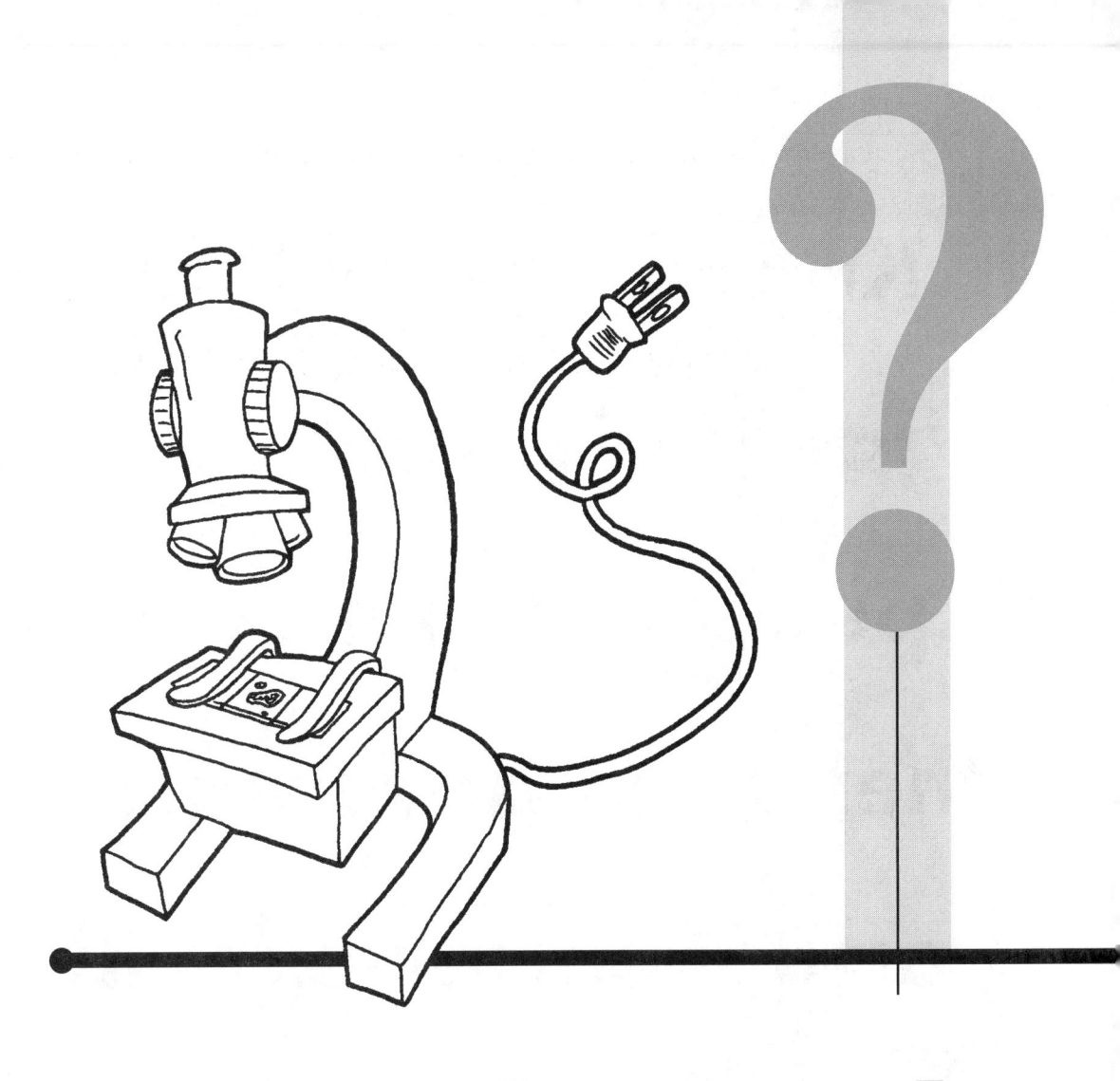

Methods and Tools

Technology that is used to help further scientific study.

higher-level thinking questions

> **Knowledge will forever govern ignorance: and a people who mean to be their own governors, must arm themselves with the power which knowledge gives.**

— James Madison

Methods and Tools
Question Cards

Methods and Tools

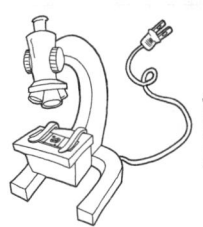

1. If you were able to talk to Anton van Leeuwenhoek, the inventor of the light microscope, how would you describe to him the impact that his invention has had on science?

Methods and Tools

2. If you were a biologist working for a zoo, what might be your most important tool? Why?

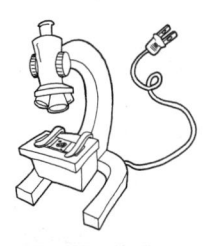

Methods and Tools

3. What are some possible uses for computers in medicine?

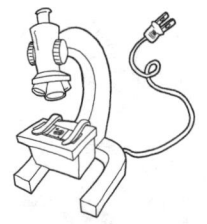

Methods and Tools

4. If you could create a new biological tool that could do anything, what would it be? How would it work?

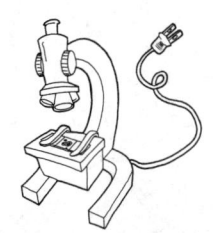

Higher-Level Thinking Questions for Biology
Kagan Publishing • 1 (800) 933-2667 • www.KaganOnline.com

Methods and Tools
Question Cards

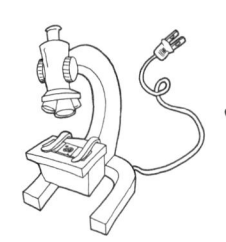

Methods and Tools

5 What impact has technology had on the field of biology?

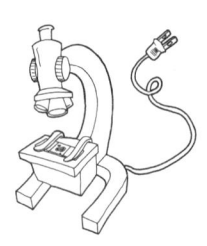

Methods and Tools

6 If you were the head of a microscope company, what technological advancements might interest you?

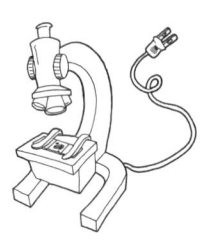

Methods and Tools

7 How has technology increased our treatment of injuries?

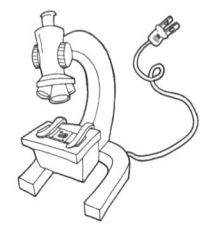

Methods and Tools

8 What do field studies and laboratory studies have in common? How do they differ?

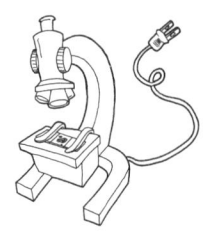

Methods and Tools
Question Cards

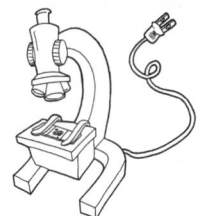

Methods and Tools

9 In what ways does biology impact your daily life?

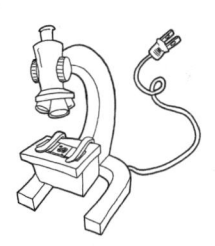

Methods and Tools

10 What is the relationship between the techniques of chromatography and gel electrophoresis?

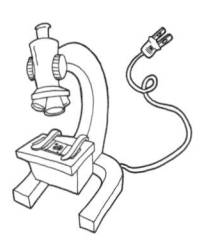

Methods and Tools

11 If you could ask a question of any scientist, who would it be? What would you ask him or her?

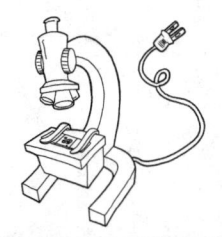

Methods and Tools

12 What are the similarities and differences between observing a specimen with a light microscope and an electron microscope?

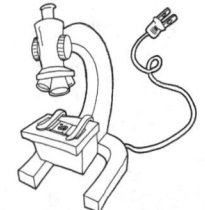

Methods and Tools
Question Cards

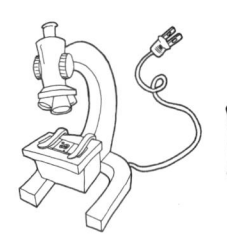

Methods and Tools

13 How would the world of biology be different if we did not have the use of microscopic technologies?

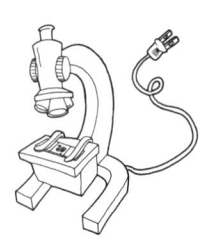

Methods and Tools

14 How could you explain the technique of performing a controlled experiment? Why are controlled experiments important in science?

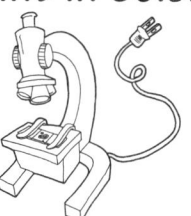

Methods and Tools

15 What is the difference between an observation and an inference? How are they alike?

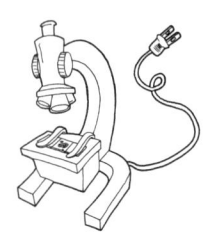

Methods and Tools

16 How might you apply the scientific method to your own real-life situations?

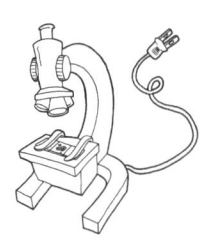

Methods and Tools
Journal Writing Question

Write your response to the question below.
Be ready to share your response.

How would it have felt to be the first biologist to have sequenced a human gene? Write a first-person account of your feelings about your scientific discovery.

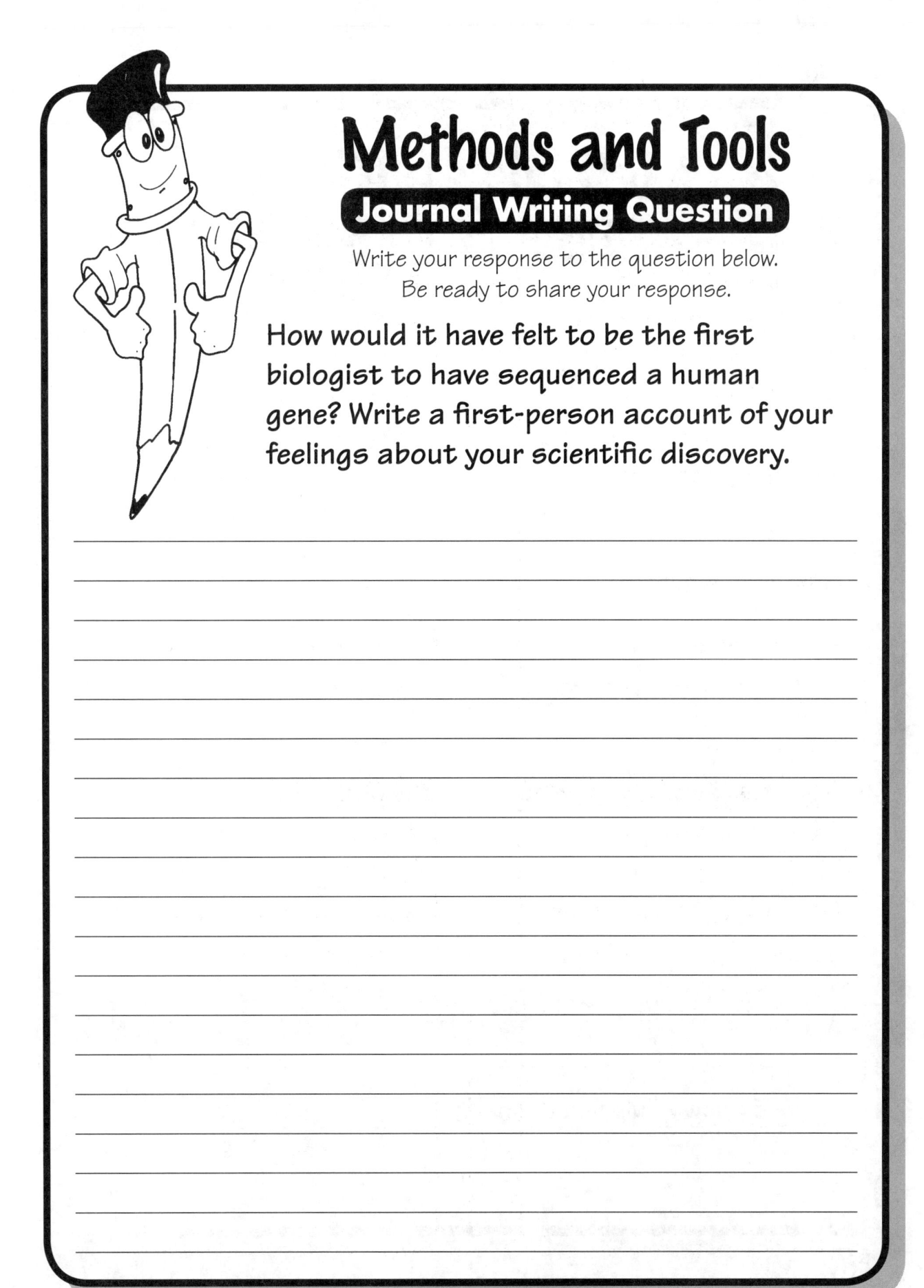

Methods and Tools
Question Starters

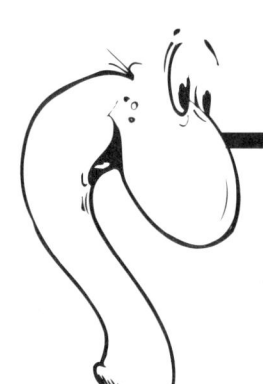

Use the question starters below to create complete questions. Send your questions to a partner or to another team to answer.

1. What impact might microscopes have

2. Why do scientists

3. How do computers

4. Do technological advancements in medicine

5. What tools would you use to solve the problem of

6. How could you test

7. What experiment could you perform to test for

8. How would you define

Monera

The kingdom of life that includes bacteria — the prokaryotes.

higher-level thinking questions

> "A prudent question is one half of wisdom."
>
> — Francis Bacon

Monera
Question Cards

Monera

1 If you were a Pool Manager, maintaining the cleanliness of a pool, how do you think Kingdom Monera would pertain to you?

Monera

2 Do you think it is possible for a single-celled moneran to completely eradicate an entire species? Explain your answer.

Monera

3 What are the similarities and differences between the archaebacteria and eubacteria?

Monera

4 In your own words, what are the characteristics that make monerans so different from all other life on Earth?

Monera
Question Cards

Monera

5 How would you describe the steps of bacterial division to a blind person?

Monera

6 Why is it so important that people take all of the antibiotics prescribed by a doctor?

Monera

7 Do you think it is a good idea for governments to keep stockpiles of deadly infectious bacteria? Explain your answer.

Monera

8 What do you think would happen if all monerans became extinct?

Monera
Question Cards

Monera

9 Imagine you are a moneran who has just entered your host organism. Describe your new life.

Monera

10 What are some of the benefits and drawbacks of using antibacterial soap in your bathroom?

Monera

11 If the nitrogen-fixing bacteria on Earth were suddenly destroyed, what would happen to other life-forms?

Monera

12 How could you design an experiment to determine which area has more bacteria, your school cafeteria or your kitchen at home?

Higher-Level Thinking Questions for Biology
Kagan Publishing • 1 (800) 933-2667 • www.KaganOnline.com

Monera
Question Cards

Monera

13 Why do antibiotics only work on bacterial infections (and not on viral infections)?

Monera

14 How are bacteria both helpful and harmful in your life?

Monera

15 A vaccination for a fatal bacterial infection has become available, but only in short supply. Who should get the vaccine? Why?

Monera

16 How could you design an experiment to determine if a disease is contagious?

Monera

Journal Writing Question

Write your response to the question below.
Be ready to share your response.

Imagine you are Alexander Fleming and have just accidentally discovered penicillin. Write an article for a scientific journal describing the importance of your new discovery.

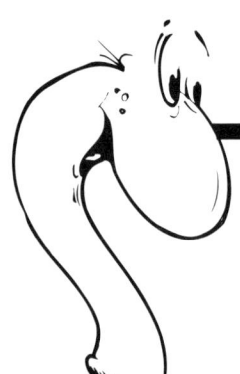

Monera
Question Starters

Use the question starters below to create complete questions.
Send your questions to a partner or to another team to answer.

1. Why is streptococcus

2. What do all monerans

3. How can a bacteria

4. If you were a bacteria

5. How is a moneran like a

6. How would you define

7. What is the significance of

8. What is the relationship between

Plants

The Kingdom Plantae — multicellular photosynthetic autotrophs that have cell walls containing cellulose.

higher-level thinking questions

> The smart ones ask when they don't know. And, sometimes when they do.
>
> — Malcolm Forbes

Plants
Question Cards

Plants

1 What features make plants so different from algae?

Plants

2 What impact do plants have on your daily life?

Plants

3 If you could be any type of plant, what kind would you choose? Why?

Plants

4 If you could extract the gene for any plant characteristic and put it into humans, which one would you choose? Why?

Higher-Level Thinking Questions for Biology
Kagan Publishing • 1 (800) 933-2667 • www.KaganOnline.com

Plants
Question Cards

Plants

5 What do you think would happen if plants didn't make sugar, but instead made only lipids through photosynthesis?

Plants

6 How might plants have evolved differently if there were half the normal amount of sunlight on Earth?

Plants

7 In one sentence, how could you describe alternation of generations?

Plants

8 Why is it so important that many different types of photosynthetic pigments exist in the Plant kingdom?

Plants
Question Cards

Plants

9 Why were cones such a significant adaptation for the Plant kingdom?

Plants

10 What are the major advantages of being a member of the Division Anthophyta?

Plants

11 Explain the biology behind the phrase, "one bad apple can spoil the bunch."

Plants

12 Design an experiment to test which wavelengths of light are required for a plant to maximize its rates of photosynthesis.

Plants
Question Cards

Plants

13 What conditions existed on early Earth that prohibited plants from growing more than a few centimeters in height?

Plants

14 Using movement, act out the steps in seed and early plant development.

Plants

15 Predict what would happen to a leaf if the guard cells re-fused to the stomata.

Plants

16 Compare and contrast the adaptations necessary for survival between a desert plant and an aquatic plant.

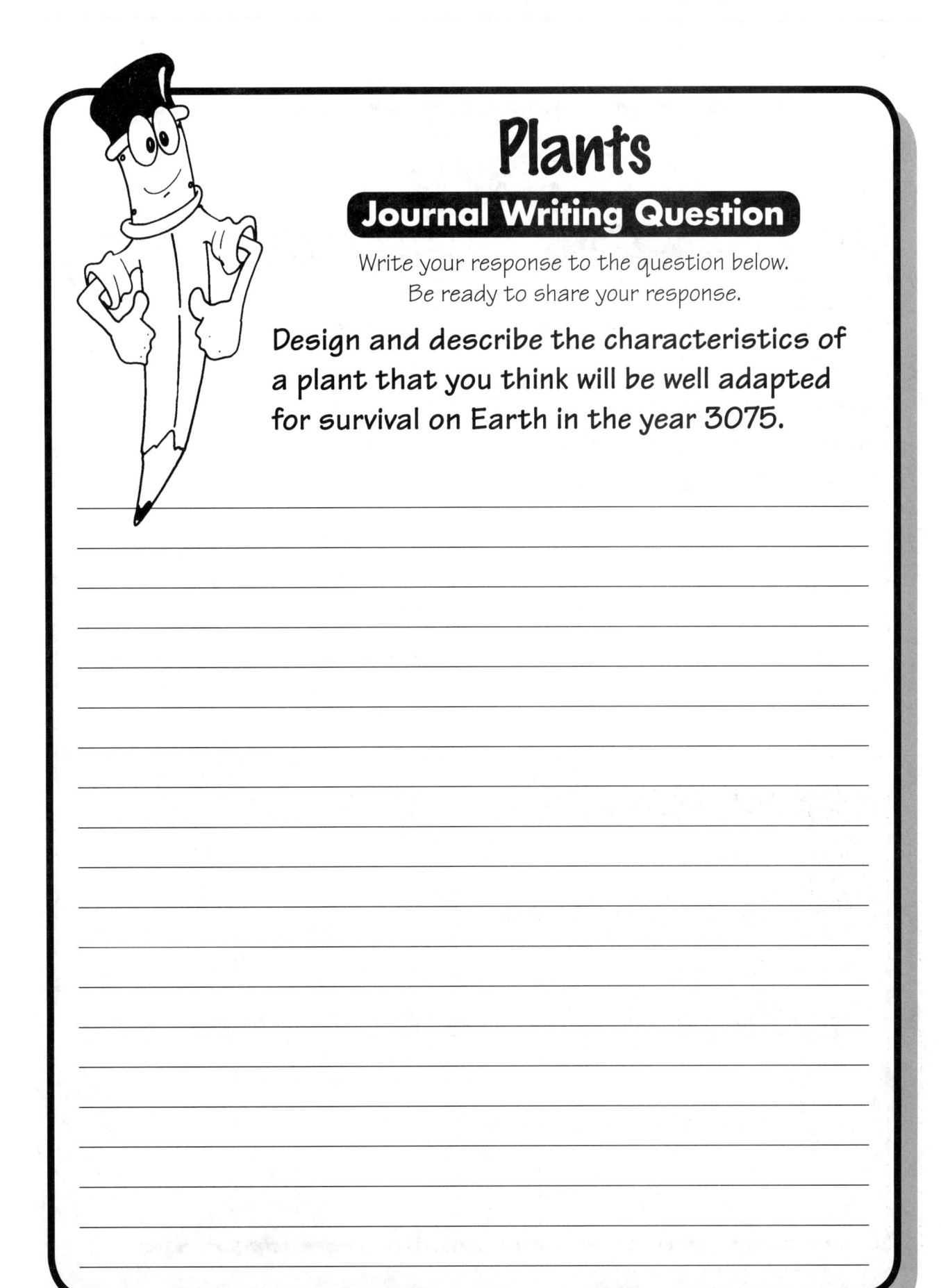

Plants

Journal Writing Question

Write your response to the question below.
Be ready to share your response.

Design and describe the characteristics of a plant that you think will be well adapted for survival on Earth in the year 3075.

Plants
Question Starters

Use the question starters below to create complete questions.
Send your questions to a partner or to another team to answer.

1. How can plants

2. What part of a plant might

3. How could you use plants for

4. Why are mosses

5. What would it feel like to be a

6. How are carnivorous plants like

7. How can you explain why plants

8. What impact might ferns

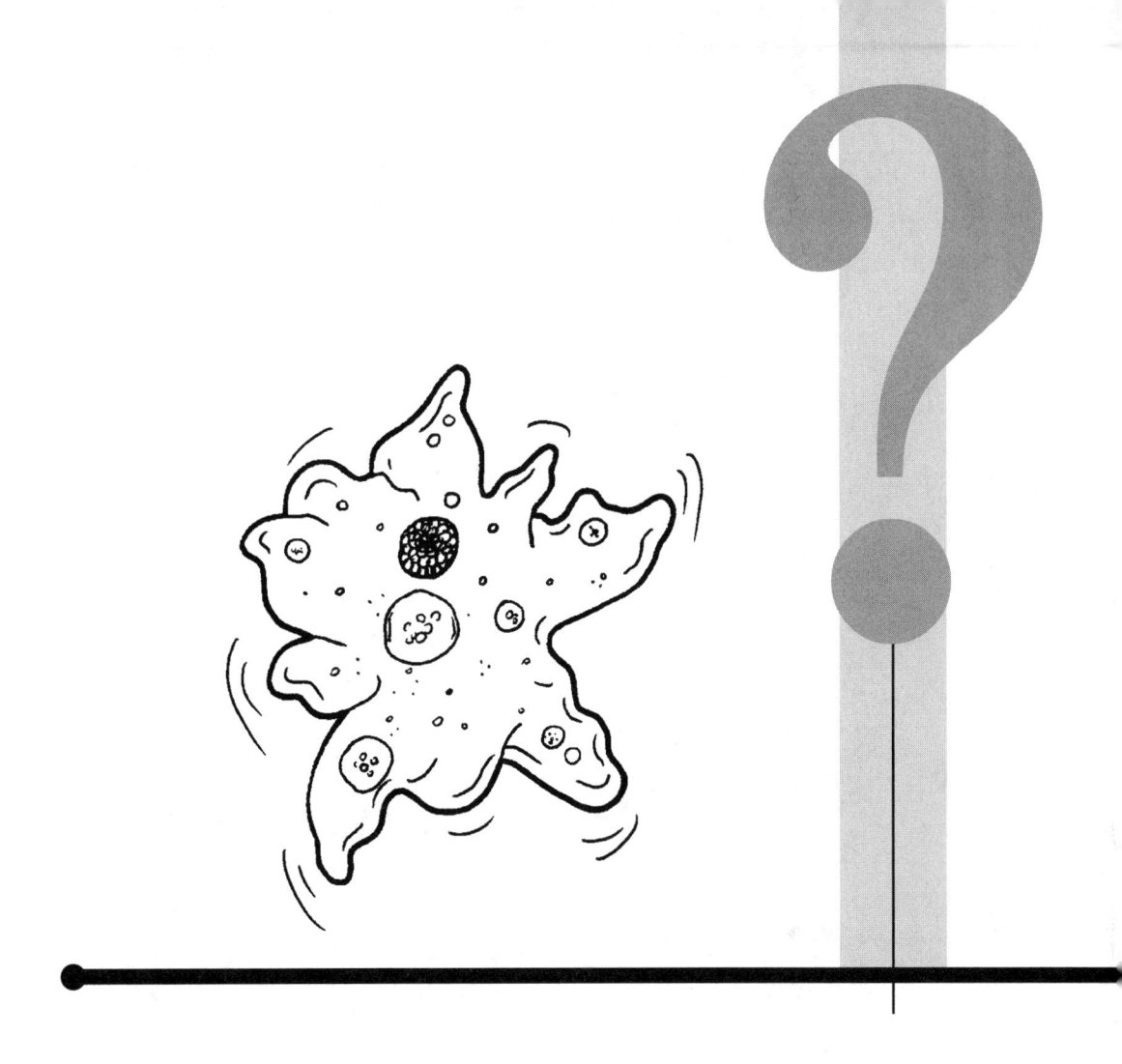

Protista

The kingdom of life that is characterized as eukaryotic cells that do not belong to the kingdoms Plantae, Animalia, or Fungi.

higher-level thinking questions

> **Never stop questioning.**
> — Albert Einstein

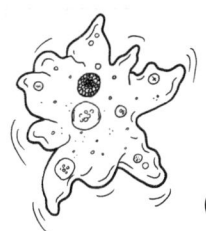

Protista
Question Cards

Protista

1 Why is Kingdom Protista considered to be a stepping stone to the three other eukaryotic kingdoms?

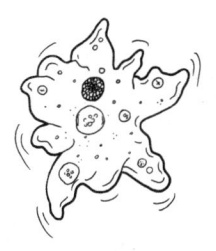

Protista

2 Why do you think the members of Division Chlorophyta are thought to be the descendents of modern plants?

Protista

3 What characteristics do animal-like protists have that make them similar to animals?

Protista

4 How are plant-like protists different from true plants?

Higher-Level Thinking Questions for Biology
Kagan Publishing • 1 (800) 933-2667 • www.KaganOnline.com

Protista
Question Cards

Protista

5 How are fungus-like protists similar to fungi?

Protista

6 How could you summarize the link between protista and bacteria?

Protista

7 What is plankton and why is it so important in an ecosystem?

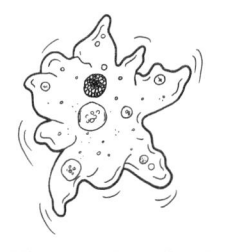

Protista

8 Do you think that it would be possible for a land organism to have both characteristics of plants and animals like a euglena?

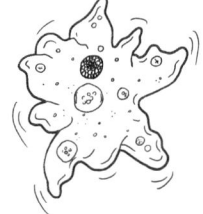

Protista
Question Cards

Protista

9 Create a public service announcement that explains the risk of drinking from unapproved water sources.

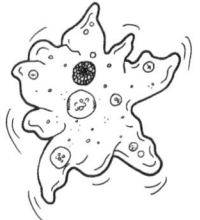

Protista

10 Compare and contrast a human sperm cell with a zooflagellate.

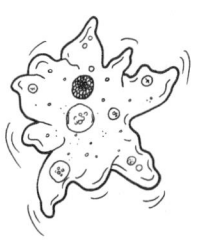

Protista

11 What relationship can be drawn between the rates of malarial infection (caused by a protist) and the number of people with hemophilia in different areas of the world?

Protista

12 How could you design an experiment that manipulates environmental factors to determine the efficiency of the contractile vacuole's action within a paramecium?

Protista
Question Cards

Protista

13 What do you predict would happen to termites if the protists inhabiting their guts were suddenly killed?

Protista

14 What do you predict would happen to the life-forms in a small body of self-contained water that was overtaken by a mass of green algae?

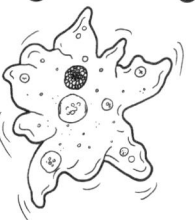

Protista

15 Describe what it would have been like to be the first cell taken in by endosymbiosis. Discuss the changes as you adapted to begin functioning as the cell's nucleus.

Protista

16 What general rule could be applied to all protists to justify their classification under one kingdom?

Protista

Journal Writing Question

Write your response to the question below.
Be ready to share your response.

Write an editorial for a scientific journal that defends your position of reclassifying the green algae as members of the Plant kingdom.

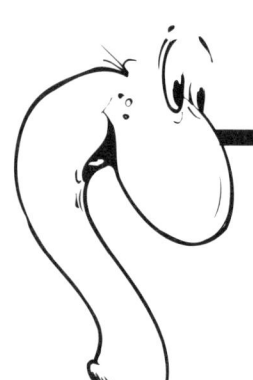

Protista
Question Starters

Use the question starters below to create complete questions. Send your questions to a partner or to another team to answer.

1. If you were a paramecium

2. Why is it so important that dinoflagellates

3. What impact might slime molds have on

4. What do you think would happen if Division Chlorophyta

5. If you were a

6. How can you express in a drawing

7. How could you break down

8. What examples

Viruses and Diseases

The study of particles that impact living organisms.

higher-level thinking questions

> **Discovery consists in seeing what everyone else has seen and thinking what no one else has thought.**
>
> — Albert Szent-Gyorgi

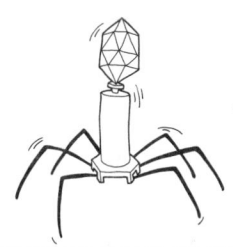

Viruses and Diseases
Question Cards

Viruses and Diseases

1 In your opinion, is a virus living or nonliving? Explain your answer.

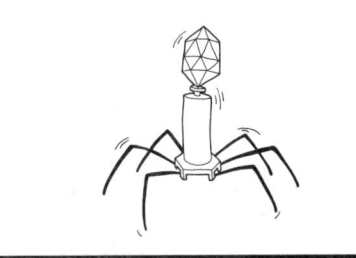

Viruses and Diseases

2 How does our immune system function as if it has a memory?

Viruses and Diseases

3 If you could create a drug to cure any one disease in the world, which disease would you choose? Why?

Viruses and Diseases

4 What would be the impact if all antibiotics we currently have suddenly stopped working?

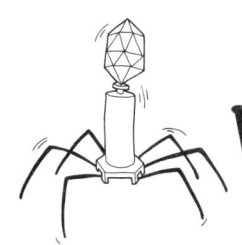

Viruses and Diseases
Question Cards

Viruses and Diseases

5 Why is pest control so important to human disease prevention?

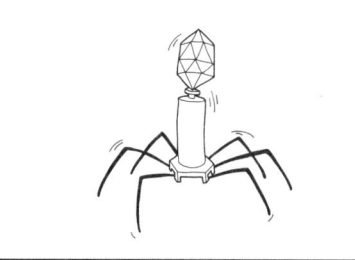

Viruses and Diseases

6 What benefits could stem cell research have in fighting diseases?

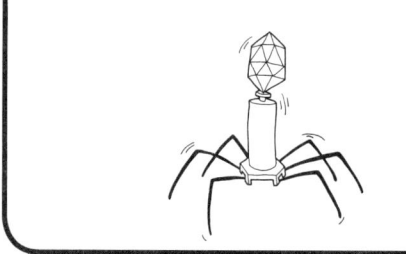

Viruses and Diseases

7 What risks could modern medical technology create in curing diseases?

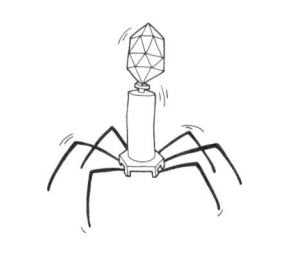

Viruses and Diseases

8 What are some ways we could help reduce the spread of HIV/AIDS? Explain.

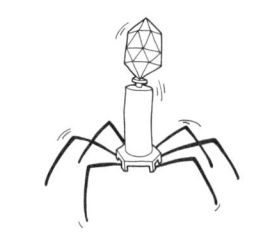

Viruses and Diseases
Question Cards

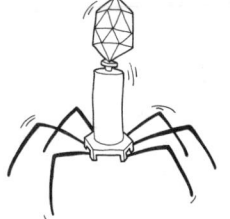

Viruses and Diseases

9 Should vaccinations be required in order to attend public school? Explain your views.

Viruses and Diseases

10 Why, when a virus has a new "outbreak" is it so traumatizing so quickly in the 21st century?

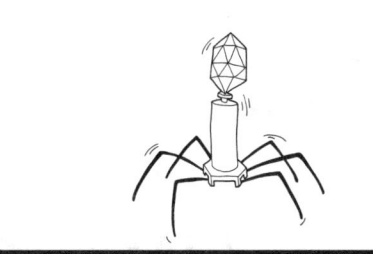

Viruses and Diseases

11 Is it safe to say that viruses are the new weapons? Why or why not?

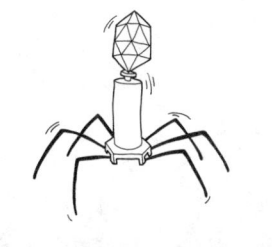

Viruses and Diseases

12 What are the characteristics of a virus that make them so difficult to detect and treat?

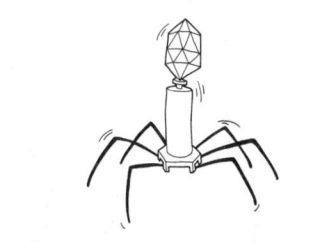

Higher-Level Thinking Questions for Biology
Kagan Publishing • 1 (800) 933-2667 • www.KaganOnline.com

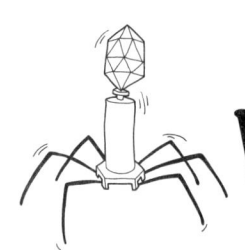

Viruses and Diseases
Question Cards

Viruses and Diseases

13 "The single biggest threat to man's continued dominance on the planet is the virus." Joshua Lederberg PhD, Nobel Laureate
Do you agree or disagree? Why?

Viruses and Diseases

14 How can the views of society impact research on new disease?

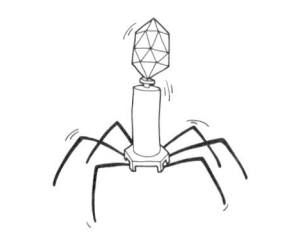

Viruses and Diseases

15 You are appointed to determine if the risks of a vaccination are worth the prevention of a disease. What are your conclusions?

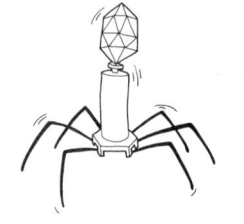

Viruses and Diseases

16 How is a virus different from a disease? How is it similar?

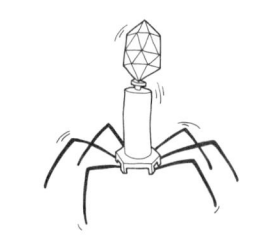

Viruses and Diseases
Journal Writing Question

Write your response to the question below.
Be ready to share your response.

You are in charge of developing a public health initiative to reduce the number of new HIV infections in sub-Saharan Africa. What steps would you take to help prevent the further spread of HIV?

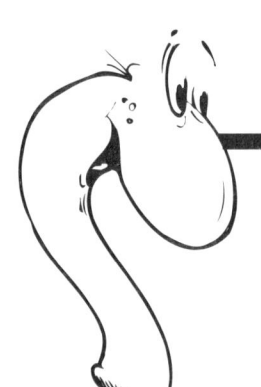

Viruses and Diseases
Question Starters

Use the question starters below to create complete questions.
Send your questions to a partner or to another team to answer.

1. What would happen if malaria

2. When do you think influenza will

3. Do you think bacteria

4. Why do viruses

5. Should genetic diseases

6. How could you get rid of

7. What preparation would you take

8. In your life, how can you apply

Question Books!

Light the Fires of Your Students' Minds with this Terrific Series of Higher-Level Thinking Question Books!

Promote Non-Stop Discussion

Sharpen Thinking Skills

Improve Writing Skills

Loaded with Hundreds of Provocative, Intriguing, Mind-Stretching Questions and Activities!

Call for Free Catalogs! Or Visit Us Online!

1 (800) 933-2667 **Kagan** www.KaganOnline.com

Kagan

It's All About Engagement!

**Kagan is your source
for active engagement in the classroom.**

Check out Kagan's line of books, SmartCards, software, electronics, and hands-on learning resources—all designed to boost engagement in your classroom.

Books

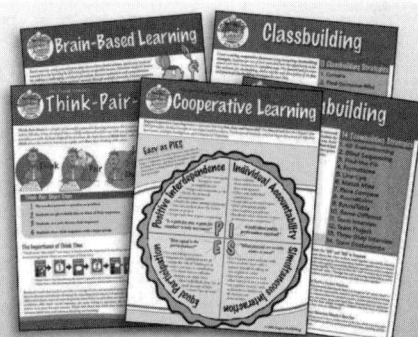

SmartCards

Spinners

Learning Chips

Posters

Learning Cubes

KAGAN PUBLISHING

www.KaganOnline.com ★ 1(800) 933-2667

Kagan

It's All About Engagement!

Kagan is the world leader in creating active engagement in the classroom. Learn how to engage your students and you will boost achievement, prevent discipline problems, and make learning more fun and meaningful. Come join Kagan for a workshop or call Kagan to **set up a workshop for your school or district**. Experience the power of a Kagan workshop. **Experience the engagement!**

SPECIALIZING IN:

- ★ Cooperative Learning
- ★ Win-Win Discipline
- ★ Brain-Friendly Teaching
- ★ Multiple Intelligences
- ★ Thinking Skills
- ★ Kagan Coaching

KAGAN PROFESSIONAL DEVELOPMENT

www.KaganOnline.com ★ 1(800) 266-7576